THE TWIN WHO WASN'T

Chimerism and the Blurred Boundaries of Genetic Identity

RAFAEL OYELARAN

This book is a work of nonfiction. The information presented is based on publicly available research, documented case studies, and scientific literature current as of the time of writing. It is intended for general informational and educational purposes only and does not constitute medical, legal, or professional advice. Readers with specific medical or legal concerns should consult a qualified professional.

Every effort has been made to ensure the accuracy of the information contained in this book at the time of publication. The author assumes no responsibility for errors, inaccuracies, omissions, or any inconsistency herein.

TABLE OF CONTENTS

CHAPTER 1

The Test That Said She Wasn't Her Own Mother

In the autumn of 2002, in a government office in Washington State, a caseworker slid a piece of paper across a desk toward a twenty six year old mother of two named Lydia Fairchild. Lydia was applying for public assistance. She was pregnant with her third child, recently separated from the children's father, and doing what millions of parents in similar circumstances do every year: filling out routine paperwork so the state could pursue child support on her behalf. The process required a DNA test, a formality so common it rarely earns a second thought. Cheeks were swabbed. Samples were sent to a lab. Lydia went home and waited for a result she assumed would be a rubber stamp.

The result that came back was not a rubber stamp. It was an accusation.

The DNA test confirmed, beyond any statistical doubt, that the children's father was who everyone already knew him to be. But it delivered a second finding that upended everything: Lydia Fairchild, the woman who had carried these children for nine months each, who had felt them kick, who had pushed them into the world in a hospital delivery room, who fed them and bathed them and knew the sound of their individual cries in the dark, was not, according to the test, their biological mother.

There was no ambiguity in the numbers. A maternity test does not deal in probabilities the way some forensic comparisons do. It either matches or it does not, and this one did not. Somewhere in a lab, a technician looked at a printout that said, in the flat and merciless language of genetics, that the woman who said she was the mother of these two children shared no detectable genetic relationship with either of them.

The state's response was swift and severe. If Lydia was not the biological mother, then one of two things had to be true. Either she was committing welfare fraud by claiming children who were not hers, perhaps as part of a surrogacy arrangement for someone else, or she had somehow acquired these children through means the state needed to investigate. Prosecutors moved to have her children removed from her custody. A court date was set. Lydia, who had never doubted for one second that she was her children's mother, found herself facing the very real possibility of losing them because a laboratory test said something that her entire lived experience insisted could not be true.

This is where most stories about bureaucratic error end, with either a clerical mistake discovered and quietly corrected, or a tragedy that unfolds because nobody looked closely enough. Lydia's story went somewhere else entirely, into territory that would eventually force geneticists, lawyers, and journalists to reconsider what they thought they understood about the reliability of DNA evidence, and what it means, at the most fundamental biological level, to be a single person.

Lydia's case worker did not accuse her lightly. DNA testing in family court has, for decades, carried an aura of scientific infallibility. It is presented to juries and judges as close to unimpeachable, a technology that cuts through the messiness of human testimony and delivers an answer rooted in the actual molecular truth of who belongs to whom. In the popular imagination, DNA evidence exonerates the wrongly convicted and identifies the guilty with a certainty that eyewitnesses and fingerprints cannot match. It made sense, from the state's perspective, that if two people's DNA did not match, then no biological relationship existed between them. That assumption held for millions of maternity and paternity cases every year. It simply did not hold for Lydia.

Lydia hired an attorney named Alan Tindell, a man who would spend the next several years of his career on a single, strange, and increasingly consuming case. Tindell's first instinct was to have the test redone. Labs make mistakes. Samples get mislabeled, cross contaminated, or degraded. A second test, run independently, could settle the matter quickly if the first result had simply been an error.

The second test came back and told the same story as the first. No match.

At this point the case moved from strange to alarming. A court hearing loomed in which the state intended to argue that Lydia's children should be removed from her home and that she should face potential fraud charges. Tindell needed something more than a request for a retest. He needed an explanation for how a woman who had unquestionably given birth to two children, in hospitals, with medical staff present, could fail a genetic maternity test not once but twice.

He found part of his answer by accident, in a case unfolding roughly three thousand miles away on the other side of the country, in Boston.

A woman named Karen Keegan needed a kidney transplant. Her three adult sons, hoping to be tested as potential donors, went through the standard genetic compatibility screening that precedes any living organ donation. The results were bewildering. According to the DNA analysis, two of Karen's three sons appeared to share no genetic relationship with her at all. Only one showed a match consistent with an ordinary mother and son relationship.

Karen, like Lydia, had never doubted that she was the biological mother of all three of her children. She had given birth to each of them. There was no scenario in her own memory or her family's history that could explain a genetic test insisting otherwise for two thirds of her sons.

Doctors at Massachusetts General Hospital, faced with a medical puzzle rather than a legal one, had the luxury of curiosity that a courtroom rarely affords. They began testing tissue from different parts of Karen's body, rather than relying on a single blood sample. Blood tests continued to produce the confusing mismatch. But when researchers examined tissue from a thyroid nodule that had been surgically removed from Karen years earlier, they found something remarkable. That tissue carried a different genetic signature than her blood, a signature that did, in fact, match her sons.

Karen Keegan's body was not one genetic individual. It was two, coexisting within a single physical form, distributed unevenly across different organs and tissues. Her blood carried the DNA of one genetic line. Her thyroid, and as it turned out other tissues as well, carried the DNA of another. She had, at some point in her earliest embryonic development, absorbed the cells of a twin who had never developed into a separate, viable being. The condition responsible for this was named decades earlier by researchers who had encountered similarly bizarre cases: chimerism, borrowed from the mythological chimera, a creature stitched together from the parts of different animals.

Karen's case had been documented and published in a respected medical journal, giving Tindell exactly the precedent he needed. If chimerism could explain how a woman's blood might fail to match her own biological children, the same mechanism could explain Lydia's situation. The question now was whether Lydia herself was a chimera, and if so, where in her body the matching genetic signature might be hiding.

This was easier to theorize than to prove. Chimerism is not a condition doctors typically go looking for, because there is rarely a reason to suspect it. Lydia's hair, saliva, and blood had all already been tested, and none of them matched her children. If she was a chimera, the genetic line that had produced her eggs, and therefore her children, was hiding somewhere else in her body, in tissue nobody had thought to sample.

The court, persuaded by the Keegan precedent that a legitimate biological explanation might exist, agreed to allow further testing before making a final custody determination. This is itself worth pausing on. A courtroom, an institution built on precedent and established fact, was being asked to entertain a genetic phenomenon so rare that most working physicians would go an entire career without encountering a single confirmed case. Yet the alternative, ordering a mother separated from her children on the strength of a DNA test that everyone increasingly suspected might be technically accurate and substantively misleading, was worse.

Investigators eventually turned to a tissue type that had not yet been examined: a cervical smear. When Lydia's cervical cells were tested, the genetic profile matched her children. Her reproductive tissue, the cells that had actually produced the eggs fertilized to create her children, carried a different genetic signature than her blood and her skin. She was, like Karen Keegan, a chimera, two genetically distinct cell lines coexisting in one body since the earliest days of her own embryonic development, likely the result of two fraternal, non identical twin embryos that fused together and continued developing as a single individual.

The charges against Lydia were dropped. She retained custody of her children. The bureaucratic machinery that had nearly torn her family apart quietly reversed course, and life, for Lydia Fairchild personally, moved forward from there with far less drama than her case had generated in legal and scientific circles.

But the story did not end with her custody being restored. It could not, because what her case revealed was too significant to file away as a resolved anomaly. Lydia Fairchild's ordeal became one of the most widely cited examples in the world of a phenomenon that challenges something most people never think to question: the assumption that every human body contains exactly one set of genetic instructions, uniform across every cell, from scalp to fingertip to reproductive organ.

That assumption is wrong far more often than most of us realize. Not wildly wrong, not commonly wrong in ways that produce headline making court cases, but wrong at a low background rate that researchers are only beginning to understand. The mechanisms that produced Lydia's condition and Karen's condition, the quiet absorption of one early embryo by another, may be involved in a substantial fraction of all human pregnancies that begin as twins, most of which nobody ever knows occurred.

This book exists because of cases like Lydia Fairchild's and Karen Keegan's, and because of the broader scientific and philosophical territory they open up. It is a book about what happens when the twin who was supposed to exist alongside you does not survive the earliest weeks of gestation, but does not entirely disappear either. It is about blood chimerism between twins who both survive pregnancy, sharing a placenta and, occasionally, a permanent exchange of blood forming cells. It is about the microchimerism that occurs during every single pregnancy, when fetal cells cross the placental barrier and take up long term residence in a mother's body, sometimes for decades, and when maternal cells do the same in reverse. It is about the legal system's uneasy relationship with genetic evidence once that evidence turns out to be more complicated than anyone assumed. It is about the philosophical unease that chimerism provokes once you start asking what, precisely, makes a person a single, continuous, bounded individual in the first place.

Most of all, it is about a form of biological reality that sits far closer to ordinary human experience than the rarity of confirmed diagnoses would suggest. Fewer than a hundred cases of tetragametic chimerism, the type found in Lydia and Karen, have been definitively documented worldwide. But researchers who study the phenomenon suspect that number reflects how rarely anyone looks for chimerism, not how rarely it occurs. Most chimeras almost certainly go through life never being tested in a way that could reveal their condition, because there is ordinarily no reason to test. A person whose blood and skin and organs all happen to derive predominantly from the same one of two original cell lines will show no outward sign whatsoever. Lydia Fairchild was only discovered because a government bureaucracy demanded a DNA test for an unrelated administrative purpose, and because her particular distribution of the two cell lines happened to place the mismatching lineage in her reproductive tissue rather than somewhere less consequential.

Somewhere in the world, quite possibly among people reading this very page, there are individuals who are, biologically speaking, the fused remnant of two people who were never meant to become one. They do not know it. There is usually no reason for them to find out. And yet the question of what that means, for identity, for inheritance, for the stories we tell about where we come from, does not require a diagnosis to be worth asking.

The chapters that follow will take you through the science of how chimerism happens, the individual human stories of people who discovered they carried it, the courtroom dramas it has produced, the philosophical puzzles it forces us to confront, and the surprisingly ancient mythological tradition that gave the condition its name centuries before anyone understood its biological mechanism. Along the way, you will meet a model whose torso is visibly divided into two distinct skin tones by the twin she absorbed before birth. You will encounter forensic scientists forced to explain to bewildered juries why a rapist's DNA appeared to belong to two different men. You will learn why every mother who has ever been pregnant carries, quite literally, a small population of her child's cells inside her own body for years afterward, and why that fact is reshaping how doctors think about pregnancy, autoimmune disease, and the biological bond between parent and offspring.

Lydia Fairchild's case forced open a door that had been sitting ajar in the scientific literature for decades. This book walks through it.

It is worth noting, before moving forward, exactly how narrowly Lydia's case turned on a piece of scientific documentation that happened to exist at precisely the moment it was needed. Karen Keegan's case, examined in full in a later chapter, had been published only a few years before Lydia's ordeal began, and had it not been published, or had Lydia's attorney simply failed to encounter it during his research, there is no guarantee the court would have granted the additional testing that ultimately vindicated her. Legal proceedings operate under real time pressure, with hearing dates fixed well in advance and judges required to reach decisions based on the evidence presented to them within that fixed window. A mother in Lydia's position, absent the fortunate timing of Karen Keegan's published case, might easily have exhausted her legal options before any alternative explanation for her impossible test results could be identified and confirmed, a sobering thought that underscores just how much of Lydia Fairchild's ultimate vindication depended on circumstances beyond her own control or her attorney's diligence alone.

Alan Tindell has spoken publicly, in the years following the case, about how close the outcome actually came to going the other way, describing the period before Karen Keegan's case surfaced in his research as one of genuine uncertainty about whether he would be able to construct any scientifically credible defense at all. The state, for its part, was not acting out of malice or incompetence in pursuing the case as aggressively as it did. Welfare fraud involving surrogate pregnancies is a real phenomenon that state agencies have a legitimate interest in investigating, and a DNA test showing no genetic relationship between a woman and the children she claims as her own is, in the vast majority of cases that any caseworker will ever encounter, accurate evidence of exactly that kind of fraud. Lydia Fairchild's case was frightening precisely because the institutions involved were behaving reasonably, by the standards of what they knew and what they had reason to expect, and still came within striking distance of a profound injustice.

This is a pattern worth carrying forward into the chapters that follow, because it recurs across nearly every subsequent chimerism case this book will examine. The people and institutions confronting confusing chimerism related evidence, whether a caseworker in Washington State, a surgical team in Boston, or a forensic investigator in Seoul, were rarely behaving carelessly or unreasonably given the information available to them at the time. The problem was never simple negligence. The problem was that DNA testing had, by the early twenty first century, earned a reputation for near infallibility so strong that few people working within legal or medical institutions had any established framework for questioning a result, however strange, once it came back from an accredited laboratory. Chimerism does not indict DNA testing as a technology. It indicts the assumption, quietly built into how many institutions use that technology, that a single genetic test from a single tissue sample can always be treated as the final, unambiguous word on a biological question, without room for the rare, well documented exceptions this book exists to explore.

Lydia Fairchild's own account of the ordeal, given in subsequent interviews after the case concluded, emphasizes a detail easy to overlook amid the scientific drama of her eventual vindication: the sheer duration and emotional toll of the uncertainty itself. The legal process stretched across roughly two years from the initial test result to the final resolution confirming her chimerism and closing the case against her, two years during which Lydia lived with the genuine possibility, renewed at every court date and every new round of testing, that she might lose her children despite knowing, with complete personal certainty, that she was exactly who she had always known herself to be. This lived experience of prolonged uncertainty, sustained against a backdrop of unwavering personal conviction that outside institutions simply refused to credit without additional proof, recurs as a theme across many of the human stories this book will go on to tell, and it is worth holding onto as a counterweight to the more purely scientific and legal analysis that necessarily makes up the bulk of the chapters ahead. Behind every confirmed chimerism case examined in this book stood a real person, living through real uncertainty, for a real and often uncomfortably long stretch of time, before the underlying biological explanation was finally identified and confirmed.

CHAPTER 2
What Is a Chimera

How many people are you?

For all but a vanishingly small fraction of humanity, the honest answer is one. Every cell in your body, from the neurons firing in your brain to the skin cells flaking off your elbow, carries the same genetic blueprint, copied faithfully from the single fertilized egg that began your existence. This uniformity is so basic to how we understand biology that it rarely gets stated outright. It is simply assumed, the way we assume gravity will keep our feet on the floor. One person, one genome, present in every one of the roughly thirty seven trillion cells that make up an adult human body.

Chimerism is the exception that breaks this assumption, and understanding it requires backing up to the earliest moments of human development, before anyone would recognize the thing forming in a uterus as a person at all.

Ordinary human reproduction begins when a single sperm fertilizes a single egg. The resulting cell, called a zygote, contains a complete genetic blueprint, half from the biological father and half from the biological mother, combined into something new and genetically distinct from either parent. That zygote then begins dividing, first into two cells, then four, then eight, doubling repeatedly as it travels down the fallopian tube toward the uterus. Every one of those cells is a genetically identical copy of the original zygote. Barring mutation, which does happen at a low background rate throughout life, the DNA in your skin today is the same DNA that existed in that first zygote, copied billions of times over across billions of cell divisions.

Twins complicate this picture in one of two ways, and the distinction matters enormously for understanding chimerism.

It is worth briefly noting how twin pregnancies themselves have become considerably more common over recent decades, since this broader demographic trend forms an important backdrop for understanding how frequently the biological starting point for chimerism actually arises. Rates of twin births rose substantially across many developed countries in the closing decades of the twentieth century and the opening decades of the twenty first, a trend researchers attribute primarily to two combined factors: the increasing average age at which women give birth, since ovulation of multiple eggs during a single cycle becomes somewhat more common with advancing maternal age, and the growing use of assisted reproductive technology, particularly earlier fertility treatment protocols that involved transferring multiple embryos in a single attempt specifically to improve the odds of at least one successful pregnancy. This rise in twin pregnancy rates has, indirectly, expanded the total population of pregnancies in which the specific developmental events capable of producing chimerism, whether through embryonic fusion, shared placental blood exchange, or simple vanishing twin syndrome without fusion, have the opportunity to occur in the first place.

Identical twins, technically called monozygotic twins, begin as a single zygote that splits into two separate embryos early in development. Because both embryos originated from the same fertilized egg, they carry identical genetic material. This is why identical twins share the same blood type, the same eye color in the vast majority of cases, and DNA so similar that standard genetic testing generally cannot distinguish between them at all. They are, genetically, the same person occupying two bodies, though of course their environments, experiences, and even some patterns of gene expression diverge from the moment of the split onward.

Fraternal twins, or dizygotic twins, are a different matter entirely. They result from two separate eggs, released during the same ovulation cycle, each fertilized by a different sperm cell. Fraternal twins are, genetically, no more similar to each other than any two siblings born from separate pregnancies. They simply happen to share a uterus and a birthday. They can be different sexes. They can look nothing alike. Their only shared characteristic is timing.

Tetragametic chimerism, the type of chimerism found in Lydia Fairchild and Karen Keegan, begins as a fraternal twin pregnancy. Two separate eggs are fertilized by two separate sperm cells, producing two genetically distinct zygotes, each with its own complete and separate genetic identity, exactly as would happen with any pair of fraternal twins. Under ordinary circumstances, these two zygotes would continue developing independently, sharing a uterus but remaining biologically distinct individuals, eventually born as fraternal twin siblings.

Chimerism occurs when, instead of developing as two separate embryos, the two zygotes fuse together at a remarkably early stage, typically somewhere in the window of four to twelve days after fertilization, before either embryo has developed the specialized tissue layers that will eventually become distinct organs. The fused mass of cells continues developing, but now it is developing as a single embryo built from two genetically distinct populations of cells rather than one. As that embryo grows into a fetus and eventually a full human being, the two cell populations distribute themselves throughout the developing body in whatever pattern the process of embryonic development happens to produce. There is no governing rule that ensures an even split. One cell line might come to dominate the blood supply while the other predominates in reproductive tissue, exactly as it did in both Lydia and Karen. Or the two lines might be distributed more evenly, showing up in patches across skin, organs, and blood alike, the pattern that produced Taylor Muhl's visibly two toned torso, which later chapters will examine closely.

The resulting individual is, biologically speaking, the physical embodiment of what would have been two fraternal twins, if the fusion had not occurred. Geneticists sometimes describe a chimera as having "absorbed" a twin, though this phrasing can be misleading if taken to suggest that one twin somehow consumed or destroyed the other. What actually happens is closer to a merger. Neither embryo dies in the conventional sense. Both cell lineages survive and continue developing, just within a single combined organism rather than two separate ones. The chimera is not one twin who ate the other. The chimera is what both twins become, together, having never separated in the first place.

This is worth sitting with, because the emotional and philosophical weight of chimerism, which later chapters will explore in more depth, hinges on this exact point. A chimeric individual is not carrying dead tissue from a twin who failed to develop. They are carrying living, functioning, genetically distinct tissue that continues operating as part of their body for the entirety of their life. Karen Keegan's thyroid nodule was not a foreign growth. It was healthy thyroid tissue, doing exactly what thyroid tissue is supposed to do, that simply happened to carry a different genetic signature than her blood.

It is worth pausing here to properly credit the specific medical and scientific vocabulary this book relies on throughout, since precision in this vocabulary matters considerably to accurately understanding chimerism. A zygote refers specifically to the single cell formed immediately upon fertilization, before any cell division has occurred. An embryo refers to the developing organism during the earliest stages of gestation, generally understood to extend from implantation through roughly the eighth week of human pregnancy, after which the developing organism is instead referred to as a fetus through the remainder of gestation. Tetragametic chimerism takes its name directly from this precise developmental vocabulary combined with basic reproductive biology, with the prefix tetra referring to four, and gametic referring to gametes, the reproductive cells, meaning sperm and eggs, that combine during fertilization. A tetragametic chimera is, quite literally by name, an individual who developed from four gametes, two sperm and two eggs, rather than the two gametes, one sperm and one egg, involved in an ordinary, non chimeric pregnancy. This precise terminology, established well before Lydia Fairchild's or Karen Keegan's cases brought public attention to the condition, reflects decades of careful scientific work establishing exactly how tetragametic chimerism differs, at the most fundamental reproductive level, from every other form of human genetic variation discussed throughout this book.

The term chimera itself predates the biological phenomenon by roughly three thousand years. In Greek mythology, the Chimera was a monstrous creature described by Homer in the Iliad as a being with the body of a lion, the head of a goat rising from its back, and a serpent for a tail, capable of breathing fire. It was not simply an animal with mismatched parts stitched together for shock value.

It represented something the ancient Greek imagination found genuinely unsettling: a single creature that was simultaneously and completely several different creatures at once, not a hybrid blend but a coexistence of distinct forms within one body. When nineteenth and twentieth century biologists needed a word for organisms built from genetically distinct cell populations, the mythological chimera offered an almost perfect metaphor, and the name stuck.

The specific scientific history behind this adoption of mythological vocabulary is itself worth a brief note here, ahead of the fuller treatment mythology receives later in this book. Biologists had, well before the specific discovery of tetragametic chimerism in humans, already begun applying the term chimera within botanical research, to describe plants exhibiting a related but mechanistically distinct phenomenon in which a single plant, often produced through grafting one plant species or variety onto the rootstock of another, develops tissue containing two genetically distinct cell populations growing alongside one another within a single combined organism. This botanical usage of the term, established within plant science before the human biological phenomenon had been documented at all, provided a ready made scientific precedent that researchers working on animal and eventually human chimerism could draw upon directly once they needed vocabulary to describe an analogous, if mechanistically distinct, phenomenon occurring in animal and human bodies rather than plant tissue.

Modern genetics distinguishes chimerism carefully from a related but mechanistically different phenomenon called mosaicism, and the distinction is one of the most commonly confused points in popular discussions of both conditions. Mosaicism occurs when a single zygote, genetically uniform at fertilization, develops a mutation partway through embryonic development. That mutation is then copied into every subsequent generation of cells descended from the mutated cell, producing a patch of tissue with slightly different genetic material than the rest of the body. But critically, in mosaicism, everything traces back to one original zygote and one original set of parents. The genetic variation arose after fertilization, through mutation, not through the fusion of two separately fertilized eggs.

Chimerism, by contrast, always involves two separate fertilization events, two separate zygotes, and by extension, the genetic possibility that the two cell lines could have come from two different fathers if the mother ovulated two eggs during a cycle in which she had intercourse with more than one partner, a phenomenon called heteropaternal superfecundation. This is rare but well documented, and it means that in principle, a chimeric individual's two cell lines might not simply represent two eggs from the same biological parents. They could, in unusual circumstances, represent genetic material from two entirely different fathers, coexisting in a single fused body.

The practical consequence of all this is that a chimeric person can, and often does, test as unrelated to their own biological children if the wrong tissue is sampled, exactly as happened to Lydia Fairchild and Karen Keegan. It also means that a chimeric individual might, at least theoretically, show two different blood types depending on which cell line dominates their hematopoietic system, the tissue responsible for producing blood cells. Documented cases exist of chimeric individuals with two distinct blood types coexisting, detectable through blood typing that reveals a mixed population of red blood cells rather than a single uniform type. Some chimeras have been found to have different colored eyes, one derived from each cell line, a condition called heterochromia that has other, more common, non chimeric causes as well, but which occasionally does trace back to chimerism upon investigation.

How common is tetragametic chimerism actually? This is one of the genuinely difficult questions in the field, and honest researchers are careful to distinguish between the number of confirmed cases and the likely actual prevalence. As of the most recent surveys of the medical literature, fewer than one hundred cases of confirmed human tetragametic chimerism have been formally documented and published. That number sounds vanishingly small, and in absolute terms, it is. But researchers who study the condition are near unanimous in believing this figure dramatically understates true prevalence, for a simple reason: chimerism is essentially only ever discovered by accident. There is no routine screening test for it. Doctors do not check for chimerism during a standard physical exam, because there is generally no reason to suspect it unless something anomalous surfaces, an unexpected DNA mismatch during a paternity or maternity dispute, a failed organ transplant compatibility test, or a visible physical marker like Taylor Muhl's divided skin pigmentation.

A person whose two cell lines happen to distribute in a way that produces no visible physical sign, and who never undergoes genetic testing for any reason, could live an entire life as an undiagnosed chimera and never know it. Given that vanishing twin syndrome, the loss of one twin early in a twin pregnancy, is now understood to occur in a substantial percentage of all twin conceptions, a topic the next chapter examines in detail, some researchers have speculated that chimerism resulting from twin fusion, rather than one twin simply dying and being reabsorbed without fusion, may be considerably more common in the general population than the documented cases would suggest. Estimates vary widely and remain genuinely uncertain, but it is not considered fringe science to suggest that thousands, and quite possibly far more, undiagnosed chimeras exist among the general population.

This uncertainty is itself an important part of the story this book tells. Chimerism sits in a strange scientific position: thoroughly documented in its mechanism, biologically well understood at the cellular level, and yet almost entirely unmeasured in its true prevalence, because the only reliable way to detect it involves genetic testing that most people never have reason to undergo. It is, in a sense, hiding in plain sight, a biological reality that exists at some meaningful frequency in the human population while remaining almost entirely invisible to both medicine and the people who carry it.

It is worth dwelling a moment longer on the distinction between chimerism and mosaicism introduced earlier in this chapter, because the two conditions are frequently confused even in otherwise well informed discussions, and because mosaicism itself, while a genuinely separate phenomenon from chimerism, turns out to be considerably more common and to touch a considerably larger fraction of the population than most people realize, offering a useful point of comparison for understanding just how much quiet genetic variation exists within what people generally assume to be a single, uniform body.

Mosaicism arises when a mutation occurs in a single cell sometime after the original zygote has already begun dividing, meaning that every cell descended from that one mutated cell will carry the mutation, while every other cell in the body, descended from cells that divided without acquiring the same mutation, will not. Because this process can occur at many different points during development, and can affect many different genes, the practical consequences of mosaicism range enormously in severity, from mutations so minor and so confined to a small patch of tissue that they produce no noticeable effect whatsoever, to mutations serious enough to produce recognizable medical syndromes affecting specific organs or systems. Segmental mosaicism affecting skin pigmentation, producing patterns of unusually light or dark skin confined to specific regions of the body following distinctive lines that dermatologists have learned to recognize, is a well documented and relatively frequently encountered example, one that can superficially resemble the kind of divided skin pigmentation associated with chimerism in cases like Taylor Muhl's, despite arising through an entirely different genetic mechanism involving only one original zygote rather than two.

Researchers studying mosaicism more broadly, extending their inquiry beyond visible skin conditions into genetic variation detectable only through laboratory testing, have increasingly found that low level mosaicism of one kind or another, present at a small enough scale to produce no visible or clinically noticeable effect, may be dramatically more common throughout the general population than earlier, cruder detection methods ever revealed. Advances in sequencing technology capable of detecting genetic variation present in only a small fraction of an individual's cells, rather than requiring a mutation to be present broadly enough to show up using older, less sensitive techniques, have found that some degree of detectable somatic mosaicism, meaning mosaicism affecting non reproductive body tissue, may be closer to a near universal feature of human biology than a rare exception, with most adults carrying at least some population of cells somewhere in their body harboring a genetic variant not present in the majority of their other cells, typically as an unremarkable byproduct of the sheer number of cell divisions and corresponding opportunities for copying errors that accumulate over a lifetime of ordinary cellular turnover and aging.

This broader context matters for understanding chimerism specifically because it helps correct an intuitive but mistaken assumption that might otherwise creep into a reader's thinking after encountering the dramatic, rare cases profiled throughout this book: the assumption that genetic uniformity within a single body is the invariable rule, with chimerism representing a bizarre, almost impossible exception to an otherwise ironclad biological law. The reality, once mosaicism is properly accounted for alongside chimerism, is closer to a spectrum. At one end sits the small number of confirmed tetragametic chimerism cases like Lydia Fairchild's and Karen Keegan's, involving two entire, distinct genetic identities originating from two separately fertilized eggs. At the other end sits the ordinary, near universal background hum of minor somatic mosaicism that virtually every person carries somewhere in their body, arising from ordinary mutation rather than embryonic fusion, producing no meaningful practical consequence for the overwhelming majority of people who carry it. Chimerism, properly understood, is not really a violation of an otherwise perfectly uniform biological rule. It is one particular, relatively dramatic point along a continuum of genetic non uniformity that turns out to characterize human bodies more broadly than most people, including many practicing physicians who never specialized specifically in genetics, generally appreciate.

The chapters ahead will trace the different forms this hidden biology takes, from twins who fuse into one body before birth, to twins who remain separate but exchange blood forming cells through a shared placenta, to the universal and far more common phenomenon of microchimerism that occurs in essentially every pregnancy ever carried to term, in which mother and child exchange small populations of cells that persist for years or decades afterward. Tetragametic chimerism, the dramatic form that produced Lydia Fairchild's custody battle and Karen Keegan's failed kidney compatibility test, is the rarest and most striking version of a phenomenon that, in its gentler forms, touches nearly every family that has ever included a pregnancy.

Before turning to those gentler and more common forms, though, it is worth understanding exactly what happens to the twin that does not survive, because vanishing twin syndrome is the direct biological precursor to many of the chimerism cases documented in the medical literature, and its frequency has surprised even researchers who specialize in twin pregnancies.

CHAPTER 3

The Vanishing Twin

Somewhere between fifteen and thirty six percent of all twin pregnancies never produce twins.

That statistic, drawn from decades of accumulated obstetric research and summarized in clinical references used by physicians today, describes a phenomenon called vanishing twin syndrome, and it is one of the most quietly astonishing facts in modern reproductive medicine. It means that for every three or four women who learn, via early ultrasound, that they are carrying twins, there is a meaningful chance that only one baby will ultimately be born, not because of any dramatic medical crisis, but because one of the two embryos simply stops developing in the earliest weeks of pregnancy and is reabsorbed by the mother's body, often so early and so completely that no trace of it remains by the time of birth.

Before the widespread use of early pregnancy ultrasound, vanishing twin syndrome was essentially invisible. A woman might have conceived twins without ever knowing it, only to deliver a single healthy baby nine months later with no indication that a second embryo had ever existed. It was only with the routine use of ultrasound imaging in the first trimester, beginning in the latter decades of the twentieth century, that doctors started noticing a strange and consistent pattern: an early scan would reveal two gestational sacs, sometimes two visible embryos with detectable heartbeats, and a follow up scan weeks later would show only one. The second sac would have vanished, sometimes leaving no visible remnant at all, sometimes leaving a small flattened tissue remnant that would eventually be reabsorbed entirely or, in later losses, compressed against the placenta and identified at birth as what doctors call a fetus papyraceus, a flattened, mummified remnant of the twin who did not survive.

Medical researchers have worked to pin down exactly how common this phenomenon is, and the numbers they have produced are remarkably consistent across different studies despite variation in methodology. Clinical reviews place the overall prevalence of vanishing twin syndrome in twin pregnancies somewhere between fifteen and thirty six percent, depending on how early pregnancies are monitored and how the syndrome is defined. Pregnancies that begin with three or more gestational sacs, meaning triplets or higher order multiples, show even higher rates, with more than half of such pregnancies losing at least one of the additional fetuses before the pregnancy comes to term. Some individual studies examining triplet pregnancies have found early spontaneous reduction occurring in well over half of all cases. Pregnancies conceived through assisted reproductive technology, particularly in vitro fertilization involving the transfer of multiple embryos, show meaningfully elevated rates as well, with estimates suggesting vanishing twin syndrome occurs in roughly twenty to thirty percent of such pregnancies.

Risk factors identified across multiple studies include advanced maternal age, a greater number of embryos transferred during fertility treatment, and a higher initial number of gestational sacs detected on early ultrasound. The underlying biological cause of any individual case of vanishing twin syndrome remains, in the majority of instances, genuinely unclear. Researchers describe the etiology as largely obscure, meaning that while certain risk factors correlate with higher rates, the precise mechanism causing any particular embryo to stop developing while its twin continues is often impossible to identify after the fact. Chromosomal abnormalities in the affected embryo are suspected to play a significant role in many cases, since embryos with severe genetic anomalies are disproportionately likely to fail early in development regardless of whether they are part of a twin pregnancy or a singleton one. Placental factors, including uneven blood supply between twins sharing a placenta, are also implicated in some documented cases.

What makes vanishing twin syndrome directly relevant to the subject of this book is the question of what actually happens to the tissue of the twin that stops developing, and here the medical picture becomes considerably more complex and more directly tied to chimerism than most popular discussions of the syndrome acknowledge.

In the earliest and most common form of vanishing twin syndrome, occurring within the first several weeks of pregnancy, the affected embryo's cells are simply reabsorbed by the mother's body, or in some understanding of the process, absorbed in part by the surviving twin's developing tissue, without the two embryos ever having physically fused into a single organism. In these cases, the surviving twin is not a chimera in the tetragametic sense described in the previous chapter. The vanished twin's cells did not merge with and become a permanent, functioning part of the surviving twin's body in the way Karen Keegan's twin's cells became her thyroid tissue. The pregnancy simply reduces from two embryos to one, and life continues as an ordinary singleton pregnancy from that point forward, with the surviving child being genetically and biologically no different than any child conceived as a single pregnancy from the start.

But researchers studying chimerism have long suspected, and increasingly have direct evidence, that at least some fraction of vanishing twin cases do not follow this clean pattern of simple loss. In some cases, before one embryo definitively stops developing, the two very early embryos may partially fuse or exchange cells during the brief window in which they coexist, particularly if the loss occurs later than the very earliest days of development, when the embryos have already begun forming more differentiated tissue layers. This partial exchange can leave the surviving twin carrying a detectable, permanent population of cells derived from the twin who did not survive, cells that persist, multiply, and become integrated into specific organs or tissues of the surviving individual for the rest of that person's life.

This is precisely the mechanism many researchers believe underlies documented cases of tetragametic chimerism like Lydia Fairchild's and Karen Keegan's. Rather than a clean vanishing, in which one twin simply disappears while the other continues unaffected, some vanishing twin cases represent an incomplete or asymmetric fusion, in which one embryo's cell line comes to dominate the resulting individual's body overwhelmingly, while the other embryo's cell line persists only in isolated patches or specific organ systems. From the outside, and from any but the most searching genetic analysis, such an individual would appear to be an ordinary singleton. Internally, they would be carrying a genetically distinct, functioning remnant of a twin whose broader developmental trajectory was cut short.

This connection between vanishing twin syndrome and chimerism has significant implications for how researchers think about the true prevalence of human chimerism, a subject touched on in the previous chapter. If even a modest fraction of the millions of vanishing twin cases occurring globally each year involve some degree of cell fusion or persistent cell line exchange rather than clean and total loss, the population of undiagnosed chimeras could be substantially larger than the handful of confirmed cases in the medical literature would suggest. This remains an active area of research rather than settled scientific consensus, and responsible scientists are careful to note the significant uncertainty involved in extrapolating from documented dramatic cases to estimates of general population prevalence. But the biological plausibility of the connection is well established, and it reframes vanishing twin syndrome from a simple story of pregnancy loss into something more layered: a process that, in some unknown fraction of cases, does not so much end a twin pregnancy as fold one twin's biological legacy permanently into the body of the other.

The precise developmental window during which this partial fusion, as opposed to clean and total loss, is more or less likely to occur remains an area of ongoing scientific investigation, though researchers have proposed a general framework based on what is already well established about early embryonic development more broadly. The earliest days following fertilization, before an embryo has begun forming the specialized cell layers that will eventually become distinct organs and tissues, represent the developmental window in which the two forms of twin loss, complete separate loss versus fusion producing lasting chimerism, are most difficult to distinguish from one another after the fact, since an embryo lost at this very early stage leaves minimal physical trace regardless of whether any cellular exchange with its surviving twin occurred beforehand. As a twin pregnancy progresses further, with each embryo's cells becoming increasingly specialized and committed to specific developmental pathways, researchers generally consider full scale fusion producing tetragametic chimerism to become progressively less biologically plausible, since the differentiating tissues of each embryo have less capacity to simply merge and continue developing as a single unified structure. This general framework helps explain why tetragametic chimerism appears to specifically require fusion occurring within a narrow and very early developmental window, while later stage losses, occurring after this window has closed, are more likely either to result in the kind of complete, non chimeric loss most commonly associated with vanishing twin syndrome, or in the fetus papyraceus outcome described earlier in this chapter, in which a more developed but nonviable twin becomes compressed and preserved alongside its surviving sibling without any meaningful cellular fusion between the two.

There is also a psychological and emotional dimension to vanishing twin syndrome that deserves acknowledgment, separate from its genetic implications. Parents who learn, through early ultrasound, that they were carrying twins and then later learn that one twin has been lost, often experience a form of grief that can be difficult for others to fully appreciate or validate, precisely because so little tangible evidence of the loss exists. There was no birth, no funeral, often no physical remains to mourn in any conventional sense, and yet a real pregnancy, a real developing embryo with its own distinct genetic identity, existed and then ceased to exist. Grief counselors and researchers who study pregnancy loss have increasingly recognized vanishing twin syndrome as a legitimate source of bereavement, distinct from but related to the grief associated with miscarriage more broadly, precisely because it involves the loss of a specific, individually conceived potential child within a pregnancy that, from the outside, appears to have proceeded successfully.

For the surviving twin, when parents choose to share the fact of a vanished sibling later in life, the psychological weight can be significant as well, sometimes described by adults who learn of their vanished twin in adulthood as an unsettling reframing of their own individual identity, a discovery that they were, in the earliest days of their existence, one of two, and that the singular, unified sense of self they have always carried was not, biologically speaking, guaranteed from the very beginning. Some psychologists working with twin loss survivors describe a phenomenon they term twinless twin syndrome, encompassing a range of emotional experiences reported by people who have lost a twin, whether through vanishing twin syndrome discovered later in life, stillbirth, or death after birth, including a sometimes profound sense of incompleteness that persists despite having no conscious memory of the lost sibling whatsoever.

The question of whether and when to share information about a vanished twin with a surviving child, once parents themselves become aware of it through early pregnancy ultrasound, represents a genuinely difficult parenting decision that pediatric psychologists and family counselors have increasingly addressed as ultrasound detection of vanishing twin syndrome has become more common. There is no broad professional consensus dictating a single correct approach, and different families have reasonably arrived at different answers based on their own values and their own read of a particular child's temperament and curiosity. Some parents choose to share the information openly from an early age, treating it as simply one more true fact about their child's origin, no more fraught than sharing details about a difficult delivery or an unusual due date. Others choose to wait until a child is older and better equipped to process the information thoughtfully, reasoning that a young child may lack the conceptual framework to understand vanishing twin syndrome without unnecessary confusion or worry. Still others choose not to share the information at all, particularly in cases where the loss occurred so early in pregnancy that no meaningful biological trace persists and the information carries, in their view, limited practical relevance to their child's life going forward. Family counselors who have written on this topic generally advise that whichever approach a family chooses, the information, if and when it is shared, should be presented factually and without excessive emotional weight, mirroring the same communication principle discussed in a later chapter regarding how families should ideally approach a confirmed chimerism diagnosis more broadly.

The medical understanding of vanishing twin syndrome continues to evolve alongside improvements in prenatal imaging technology. Ultrasound resolution has improved considerably since the syndrome was first widely documented in the clinical literature, allowing earlier and more precise detection of multiple gestational sacs and, correspondingly, earlier and more precise documentation of when and how a twin pregnancy reduces to a singleton. This improved detection has, somewhat counterintuitively, increased reported rates of vanishing twin syndrome over time, not because the underlying biological phenomenon has become more common, but because doctors are now catching pregnancies at earlier stages, before natural reduction occurs, that would previously have simply been recorded as ordinary singleton pregnancies from the start.

The rise of assisted reproductive technology worldwide has added a further layer of complexity to how frequently vanishing twin syndrome is detected and documented, since fertility treatment inherently involves closer and earlier medical monitoring than the average naturally conceived pregnancy receives. A woman undergoing in vitro fertilization typically has her pregnancy confirmed and imaged considerably earlier than a woman who conceives naturally and only seeks prenatal care once pregnancy is already suspected through more traditional signs. This earlier and more consistent monitoring means fertility clinics have generated a disproportionately large share of the documented vanishing twin syndrome cases in the medical literature relative to their share of overall pregnancies, a pattern that reflects differences in detection opportunity rather than necessarily reflecting a meaningfully higher underlying biological rate of vanishing twin syndrome among assisted pregnancies specifically, though as noted earlier in this chapter, some elevated risk does appear to be genuinely associated with the higher initial embryo counts and other factors common to many fertility treatments.

Clinicians who specialize in fertility treatment have, as a direct consequence of this pattern, become particularly attentive to counseling patients about the possibility of vanishing twin syndrome before an early ultrasound is performed, specifically to prepare patients emotionally for the possibility that an initial finding of twins might not be the final outcome of the pregnancy. This counseling practice reflects a broader recognition within reproductive medicine that the earlier and more detailed information modern prenatal imaging provides, while valuable in most respects, can also introduce patients to a category of pregnancy related uncertainty and potential loss that would, in an earlier medical era relying on less sensitive detection methods, simply never have come to their attention at all. Patients who conceive without any awareness that they were briefly carrying twins are spared this particular source of anxiety and grief, an outcome that was, until relatively recently in medical history, simply the default experience for nearly every pregnancy involving vanishing twin syndrome, rather than the exception it has increasingly become as prenatal monitoring technology has advanced.

Researchers studying pregnancy outcomes have also examined whether vanishing twin syndrome carries measurable risks for the surviving twin, and the evidence here remains genuinely contested within the medical literature, described by researchers themselves as an area of ongoing controversy. Several substantial studies have found that pregnancies affected by vanishing twin syndrome carry elevated risk for preterm birth and for babies being born small for their gestational age, compared to pregnancies that were singleton from conception. One large scale analysis comparing outcomes among in vitro fertilization patients found that survivors of an early vanished twin, meaning the loss occurred within the first twelve weeks, showed similar average gestational age at birth compared to ordinary singleton pregnancies, but a measurably higher rate of preterm delivery and low birth weight outcomes. Other studies, however, have found no significant difference in perinatal outcomes between vanishing twin pregnancies and ordinary singleton pregnancies, and researchers attribute at least some of this inconsistency to varying definitions of vanishing twin syndrome across different studies, with some including later term losses that carry demonstrably higher risk due to the release of inflammatory substances from a larger, more developed twin that fails later in pregnancy, compared to very early losses that appear to carry minimal if any elevated risk for the surviving twin.

What emerges from this body of research is a picture of vanishing twin syndrome as neither a rare curiosity nor a fully understood, settled phenomenon, but rather a common event in human reproduction that medicine has only recently begun to observe directly, and whose full biological and psychological consequences researchers are still working to map. It sits at the intersection of several fields this book will continue to explore: reproductive medicine, genetics, psychology, and the philosophical questions raised whenever the boundary of individual identity turns out to be less fixed than everyday experience suggests.

Support organizations and online communities specifically dedicated to parents who have experienced vanishing twin syndrome have grown substantially in recent years, offering a further point of evidence regarding just how many families are affected by this phenomenon and how much genuine need exists for accurate information and emotional support surrounding it. These communities, largely organized independently of formal medical institutions, have played a meaningful role in normalizing conversation around a specific form of pregnancy loss that, for much of medical history, carried no name and received essentially no acknowledgment at all, since without routine early ultrasound, most families experiencing vanishing twin syndrome would simply never have learned that a twin pregnancy had ever existed in the first place. The existence and growth of these communities reflects a broader pattern common to many areas of reproductive health, in which patient advocacy and peer support networks have emerged specifically to fill gaps left by a medical system that, however scientifically capable, does not always have the resources or structure in place to provide the ongoing emotional support that a diagnosis like vanishing twin syndrome can require, particularly for families navigating grief around a loss that remains largely invisible to friends, extended family, and even some healthcare providers less familiar with the condition's psychological dimensions.

Obstetric researchers have also begun investigating whether specific counseling protocols delivered at the time a vanishing twin is first identified on ultrasound might meaningfully improve psychological outcomes for affected parents, compared to more minimal or purely clinical communication of the finding. Early research in this area, while still limited in scope, has generally supported the value of proactive, compassionate counseling delivered by trained healthcare providers at the moment a vanishing twin is identified, rather than leaving parents to process the finding entirely on their own or to seek out information and support only after independently recognizing that something meaningful had occurred. This research has contributed to a gradual shift within some obstetric practices toward treating vanishing twin syndrome as a recognized and named form of pregnancy loss deserving of the same compassionate clinical attention traditionally extended to other, more widely recognized forms of miscarriage, rather than as a minor incidental finding to be noted briefly before moving on to monitor the continuing, surviving pregnancy.

The next chapter returns to Karen Keegan's story in fuller detail, examining not just the discovery of her chimerism but the medical detective work that led to it, and what her case revealed about how thoroughly modern medicine can fail to anticipate a phenomenon that, once discovered, seems almost inevitable in retrospect.

CHAPTER 4

Karen Keegan and the Kidney That Knew

The kidneys were failing, and Karen Keegan's three sons wanted to help.

Karen, a woman in her fifties living in the Boston area, had been diagnosed with kidney disease serious enough that her doctors began discussing the eventual need for a transplant. Living donor kidney transplants, in which a healthy person donates one of their two kidneys to a family member or loved one, had become a well established and relatively safe procedure by the time Karen's case unfolded, offering significantly better long term outcomes than kidneys obtained from deceased donors. For a patient with willing and available family members, the path forward seemed straightforward. All three of Karen's adult sons volunteered to be tested as potential donors.

Living donor transplantation requires careful compatibility testing before any surgery can proceed. Blood type must be compatible, and beyond blood type, doctors examine a set of proteins called human leukocyte antigens, or HLA, which play a central role in how the immune system distinguishes the body's own cells from foreign material. A closer HLA match between donor and recipient generally translates to a lower risk of organ rejection after transplant, and children are, on average, closely related to their biological parents in HLA type, since they inherit half of their genetic material directly from each parent. Testing a patient's own biological children for compatibility is standard practice precisely because that close genetic relationship usually makes for a strong donor match.

The broader system of organ allocation and living donor evaluation within which Karen Keegan's family sought answers relies on a carefully developed set of medical protocols built up over decades of transplant medicine, protocols designed specifically to maximize the chances of a successful, long lasting transplant while minimizing risk to both donor and recipient. HLA typing represents only one part of this broader evaluation process, which also includes blood type compatibility testing, general health screening of the potential donor to ensure the procedure would not create undue risk to their own health, and psychological evaluation intended to confirm that a living donor is making a fully informed and genuinely voluntary decision to undergo a significant surgical procedure for someone else's benefit. It was within this well established, carefully designed evaluation framework, built on the assumption that a patient's biological children would show a predictable and interpretable genetic relationship to their parent, that Karen Keegan's confusing test results first surfaced, precisely because that framework had never been designed with a possibility as rare as chimerism in mind.

The test results that came back for Karen's sons made no biological sense. Two of the three sons showed genetic profiles that appeared entirely unrelated to Karen's own. Not merely a poor match, the kind of result you might expect from ordinary genetic variation between parent and child, but a profile suggesting no direct genetic relationship at all. Only the third son's results were consistent with an ordinary mother and son relationship.

For a family in the middle of a medical crisis, this was bewildering and frightening in ways that went well beyond the immediate question of finding a compatible kidney donor. Karen had given birth to all three of her sons. There was no scenario involving switched hospital records or any other mundane explanation that could account for two of three children showing no genetic relationship to their mother, particularly when the third son's results were entirely normal. If the testing was accurate, the implications reached far beyond a failed transplant match, threatening to call into question the basic biological reality of a family that had lived, for decades, with no reason whatsoever to doubt itself.

Doctors at Massachusetts General Hospital, faced with a result that defied conventional explanation, did not simply accept the mismatch and move on to other options for donor kidneys, though that would have been the path of least resistance. There was, after all, a compatible option available in Karen's husband, and the surgical team did ultimately proceed with a kidney transplant using him as the donor, a procedure that was successful. But the genetic mystery itself remained unresolved, and a research team at the hospital, recognizing that something scientifically significant was likely at play, continued investigating even after the immediate medical crisis had been addressed through other means.

The research team's approach was methodical. If Karen's blood, which is what the initial compatibility testing had sampled, showed a genetic profile inconsistent with being the biological mother of two of her three sons, the logical next step was to test other tissue types from elsewhere in her body, on the theory that if some biological explanation existed for the discrepancy, it might reveal itself through a broader sampling. Blood, saliva, and other easily accessible tissue continued to show the same puzzling profile, offering no immediate resolution. The investigation deepened.

The breakthrough came through examination of thyroid tissue. Years before the kidney crisis, Karen had undergone surgery to remove a nodule from her thyroid gland, a common and usually benign condition. That excised tissue had been preserved, as is standard practice in pathology, allowing researchers to test it directly rather than needing to obtain a fresh biopsy. When the thyroid tissue was genetically analyzed, it revealed a startling result: a genetic profile distinct from Karen's blood, and one that was consistent with being the biological mother of the two sons whose earlier tests had shown no match.

Karen Keegan's body was not genetically uniform. Some of her tissue, including her blood, carried one genetic signature. Other tissue, including at least part of her thyroid gland, carried a second, distinct genetic signature. She was, in the language that would soon be applied to her case and would later be applied to Lydia Fairchild's as well, a human chimera, an individual whose body had developed from the fusion of two genetically distinct embryos rather than a single one.

The research team's findings were significant enough to warrant publication in a major medical journal, bringing Karen Keegan's case to the attention of the broader medical and genetics research community for the first time. This publication would later prove pivotal for reasons that had nothing to do with kidney transplants and everything to do with a custody battle unfolding on the opposite side of the country. When Lydia Fairchild's attorney, searching for any precedent that might explain his client's impossible maternity test results, came across the published account of Karen Keegan's case, he found the exact scientific framework needed to argue that Lydia, too, might be a chimera, a case examined in detail in the opening chapter of this book.

Karen Keegan's case offers a valuable window into how chimerism typically comes to light, because unlike Lydia Fairchild's situation, which unfolded under the adversarial pressure of a state fraud investigation and looming custody proceedings, Karen's diagnosis emerged from a comparatively unhurried process of medical curiosity. Nobody was accusing Karen of a crime. Nobody was threatening to take her children away. Her case was, from the outset, a scientific puzzle that clinicians had the freedom to pursue methodically, testing tissue after tissue until an answer emerged. This distinction matters because it illustrates something important about how chimerism gets diagnosed in the real world: it typically requires either a strong external motivation to keep testing past an initial confusing result, as happened for Karen through ordinary medical curiosity, or an adversarial legal process forcing the issue, as happened for Lydia. In the vast majority of situations where a person's blood or another single tissue type might reveal a chimeric mismatch, no such motivation or pressure exists, and the confusing result, if it ever surfaces at all, is far more likely to be written off as a laboratory error than pursued to its actual biological explanation.

The specific mechanism producing Karen Keegan's chimerism follows the same biological pathway described in the previous chapters: two separate eggs, fertilized by two separate sperm cells during what would have been, absent fusion, an ordinary fraternal twin pregnancy, merged into a single developing embryo at an early stage of gestation. The resulting individual, Karen, carried both genetic lineages throughout her body in an uneven distribution, with one lineage predominating in her blood and skin, and the other surfacing in at least some of her thyroid tissue and, crucially, in the reproductive tissue responsible for producing the eggs that became two of her three sons.

This last point deserves particular emphasis, because it is easy to misread cases like Karen's and Lydia's as somehow calling into question whether they were truly the biological mothers of their children. They were. Karen Keegan is every bit as much the biological mother of all three of her sons as any mother who does not carry chimerism. The eggs that became her sons developed from her own ovarian tissue, fertilized by her husband's sperm, exactly as in any conventional pregnancy. What chimerism reveals is not that her motherhood was somehow partial or compromised, but that the particular tissue sampled during a compatibility test, in this case blood, happened to derive from a different one of her two cell lineages than the one that produced her eggs. Had doctors sampled her ovarian tissue directly rather than her blood from the outset, no discrepancy would ever have surfaced, because that tissue would have matched her sons perfectly.

This distinction between genetic mismatch in a sampled tissue and actual biological relationship is central to understanding chimerism correctly, and it is a distinction that popular reporting on cases like Karen's and Lydia's has not always handled with appropriate care. Sensationalized headlines describing these cases have sometimes implied that these women were somehow revealed not to be their children's "real" mothers, a framing that is not only inaccurate but does a disservice to the women themselves, who endured significant distress precisely because outside institutions questioned a biological reality that was never actually in doubt. The DNA test results were not lying about the tissue they sampled. They were accurately reporting that a specific tissue sample, blood in both cases, derived from a genetic lineage different from the one that produced these women's children. The tests were technically correct and substantively misleading, a distinction with enormous consequences when institutions like courts and state welfare agencies treat DNA results as simple, unambiguous facts rather than results requiring careful interpretation within their proper biological context.

This gap between technical accuracy and substantive meaning recurs throughout the history of forensic and diagnostic science more broadly, well beyond the specific context of chimerism, and it is worth pausing to place chimerism within that larger pattern. Fingerprint evidence, long treated by courts and popular culture alike as a form of essentially infallible identification, has in more recent decades been shown by careful researchers to carry meaningful error rates and interpretive subjectivity that earlier generations of forensic practitioners rarely acknowledged openly. Eyewitness testimony, once given enormous weight in criminal proceedings, has been shown through decades of psychological research to be considerably less reliable than courts traditionally assumed, particularly under conditions of stress or when cross racial identification is involved. DNA evidence entered forensic and legal practice riding a wave of enthusiasm specifically because it appeared to solve the reliability problems that had already been identified in these older forensic technologies, offering a form of identification grounded directly in an individual's own molecular biology rather than in the more fallible domains of human perception, memory, or pattern matching skill. Chimerism represents one of a small number of genuinely well documented circumstances in which DNA evidence itself, despite its enormous and generally well deserved advantage in reliability over these older techniques, can nonetheless produce a technically accurate result that fails to capture the full and correct biological picture, a reminder that no forensic or diagnostic technology, however powerful, is entirely free of edge cases deserving of careful, informed interpretation rather than uncritical acceptance.

Karen Keegan's case, published in the medical literature with appropriate scientific rigor, helped establish tetragametic chimerism as a documented, if extremely rare, human condition that clinicians and researchers needed to keep in mind whenever genetic testing produced results that defied conventional explanation. It became, alongside a small handful of other documented cases predating it, part of the foundational case literature that any subsequent chimerism investigation, including Lydia Fairchild's, would draw upon.

The published account of Karen Keegan's case is worth situating within the small but growing body of documented tetragametic chimerism cases that preceded and followed it, since no single case, however well documented, tells the complete story of how this condition tends to present or how it typically comes to be diagnosed. Earlier documented cases in the medical literature, dating back to the middle of the twentieth century when the concept of human chimerism was first being seriously proposed and investigated, tended to be identified through blood typing anomalies discovered incidentally during unrelated medical procedures, at a time when blood typing represented one of the few genetic tests routinely performed as part of ordinary clinical care. As genetic testing technology advanced over subsequent decades, the pathways through which chimerism came to be diagnosed diversified considerably, expanding beyond blood typing anomalies into the kind of organ transplant compatibility testing that revealed Karen Keegan's condition, the paternity and maternity testing that revealed Lydia Fairchild's, and, as later chapters will examine, an increasingly wide range of additional diagnostic pathways including forensic investigation and, more recently, incidental discovery through consumer genetic ancestry testing.

The broader medical community's response to Karen Keegan's case, once published, illustrates something important about how rare condition research actually advances. A single well documented case, rigorously investigated and published with appropriate scientific detail, can become a reference point that enables the diagnosis of subsequent cases that would otherwise have gone unexplained. Before Karen's case entered the medical literature, a physician confronted with Lydia Fairchild's confusing maternity test results might reasonably have had no framework at all for considering chimerism as an explanation, since documented human cases were so exceedingly rare. After Karen's case was published, that framework existed, ready to be applied the moment a sufficiently similar situation arose.

This pattern, in which the documentation of one exceptionally rare case creates the diagnostic pathway that allows subsequent cases to be identified and correctly understood, recurs throughout the history of rare disease research generally, and it underscores why the publication of unusual medical findings, even from a single patient, carries value that extends well beyond that one individual's care. Karen Keegan's willingness, along with her medical team's, to pursue and publish her case gave Lydia Fairchild's attorney the exact tool he needed, years later and on the opposite coast, to keep a mother from losing custody of her children.

The following chapter turns to a very different kind of chimerism case, one in which the condition did not require a legal crisis or a failed transplant compatibility test to come to light, because it announced itself, quite literally, on the surface of a woman's skin.

CHAPTER 5
The Woman Split Down the Middle

A line runs down the center of Taylor Muhl's torso, front and back, straight enough that it looks almost deliberate, as though drawn with a ruler. On one side of that line, her skin carries one tone. On the other side, a visibly different, darker tone. For most of her childhood and well into her adult life, both Taylor and the doctors who occasionally examined her believed this was simply an unusually large and sharply defined birthmark, an explanation odd enough to accept and move past, the kind of physical quirk that gets absorbed into a person's sense of their own body without ever demanding a deeper explanation.

The real explanation, when it finally arrived, reframed not just a patch of skin but Taylor's entire understanding of who and what she was.

Taylor Muhl is an American singer, songwriter, and model, born in Hollywood, California, in 1983. Unlike Lydia Fairchild and Karen Keegan, whose chimerism surfaced through DNA testing conducted for entirely unrelated legal and medical purposes, Taylor's chimerism was visible from birth, worn openly on her skin for anyone who saw her unclothed, and yet it took decades to receive its correct diagnosis, illustrating just how unfamiliar even physicians tend to be with a condition this rare.

Taylor was conceived alongside a fraternal twin sister. Early in the pregnancy, the two developing embryos fused, following the same biological process described in earlier chapters: two separately fertilized eggs merging into a single developing organism before either had progressed far enough in development to be considered anything more than an undifferentiated cluster of cells. From that point forward, a single fetus developed, carrying two genetically distinct populations of cells distributed throughout its body. In Taylor's particular case, that distribution produced an unusually clear and visible boundary, splitting her torso's skin pigmentation cleanly down the midline, front and back, with a noticeably darker tone on the left side, the side derived from her twin sister's genetic lineage.

Growing up, Taylor has described being fascinated with twins as a young child, repeatedly asking her mother whether she had a twin sibling somewhere, without any concrete reason to suspect the answer might eventually turn out to be, in a very literal and biological sense, yes. The birthmark explanation for her divided skin tone satisfied both her family and her doctors for years, in part because birthmarks covering large areas of skin, while less common than smaller ones, are not unheard of, and in part because nobody had any particular reason to suspect a rare genetic condition when a far more mundane explanation was readily available and required no further investigation.

The truth surfaced gradually, driven less by the visible skin division itself than by a pattern of confusing and persistent health problems that defied easy diagnosis. Taylor has spoken publicly about experiencing a range of autoimmune difficulties throughout her life, along with sensitivities to certain metals, foods, medications, and other substances that seemed to affect only part of her body rather than presenting uniformly. She has described having what amounts to two different immune systems operating within a single body, each associated with one of her two genetic lineages, and each reacting differently to the same external triggers. A metal that produces no reaction whatsoever on one side of her body might trigger a noticeable allergic response on the other. This kind of asymmetric, tissue specific immune reactivity is a recognized feature of chimerism in documented cases, since the immune system in any individual develops its sense of what counts as "self" during early development, and a chimeric individual's immune system may, depending on how the two cell lineages are distributed, only fully recognize one of the two lineages as entirely native, treating cells derived from the other lineage with a degree of ongoing immunological suspicion even though both lineages have been part of the same body since its earliest formation.

It was ultimately a specialist familiar with unusual dermatological and genetic cases who first raised the possibility of chimerism as an explanation for Taylor's condition, following an experience Taylor has described as a moment of genuine shock:

seeing a photograph of an infant with a strikingly similar divided skin pattern down the torso, described in the medical context in which the photograph appeared as a symptom of chimerism. Genetic testing subsequently confirmed the diagnosis, providing Taylor, well into adulthood, with the first fully coherent explanation for a physical characteristic she had carried and quietly managed since birth.

Taylor has described the emotional impact of finally receiving an accurate diagnosis as one of profound relief rather than distress, a reaction that deserves particular attention given how chimerism is sometimes portrayed in more sensationalized popular coverage as an unsettling or even frightening discovery. For Taylor, spending decades without an accurate explanation for a visible physical difference, one she had, by her own account, worked to conceal for much of her career as a model and performer out of concern about how it might be perceived, was itself the more difficult experience. Receiving the correct diagnosis, even one as rare and unfamiliar as chimerism, replaced years of uncertainty with concrete understanding, and Taylor has since become a public advocate for chimerism awareness, sharing images and information about her condition specifically to help other undiagnosed chimeras recognize potential signs of the condition in themselves.

This advocacy work is medically significant in ways that extend beyond any single individual's personal journey. Because chimerism remains so poorly understood even within much of the medical community, individuals who exhibit potential physical signs, whether divided skin pigmentation, heterochromia in which the two eyes are different colors, or other markers, may go years or decades without receiving a diagnosis simply because the physicians they see have never encountered a documented case and do not think to consider it. Public figures willing to speak openly about a rare diagnosis play a genuine role in expanding general and even professional medical awareness of conditions that might otherwise remain almost entirely confined to specialist literature that most practicing clinicians never have occasion to read.

Taylor's case also illustrates an important scientific point about the range of physical presentations chimerism can produce, extending well beyond the internal, tissue specific mismatches documented in Lydia Fairchild and Karen Keegan's cases. Because the two genetically distinct cell populations in a chimeric individual can distribute throughout the developing body in essentially unpredictable patterns, the external, visible consequences of chimerism vary enormously from one documented case to another. Some chimeric individuals show no external physical signs whatsoever, their condition detectable only through targeted genetic testing of specific tissues. Others, like Taylor, show a distinct and visually striking pattern, in her case owing to differences in melanin production between the two genetic lineages present in her skin. Documented cases in the broader medical literature include instances of heterochromia, in which each eye derives its pigmentation predominantly from a different one of the two cell lines, cases of patchy or mosaic-appearing skin pigmentation less dramatically defined than Taylor's clean midline division, and cases involving asymmetric body development, in which one side of the body is measurably larger or develops certain features differently than the other, a pattern Taylor herself has described experiencing to a modest degree.

Taylor's public advocacy has also touched on a dimension of chimerism that receives comparatively little attention in strictly clinical descriptions of the condition: the practical, everyday experience of managing a body that responds inconsistently to the outside world. She has described needing to pay unusually close attention to how different products, foods, and environmental exposures affect her, precisely because a reaction that would be predictable and consistent in a non chimeric person can, in her case, depend on which side of her body, and by extension which of her two genetic lineages, happens to be more directly exposed or more heavily involved in a given physiological response. This kind of granular, trial and error self monitoring, developed necessarily and gradually over years without the benefit of an accurate diagnosis to explain why it was needed in the first place, represents a form of practical expertise that exists almost nowhere in formal medical training, precisely because so few physicians have ever needed to develop it themselves, and precisely because chimeric patients like Taylor have often had to become their own primary experts on a condition few of their doctors have direct experience treating.

Researchers who study chimerism note that the specific pattern any individual case takes depends heavily on the precise developmental timing and location of the original embryonic fusion, factors that are essentially impossible to reconstruct after the fact for any given individual, since the fusion itself occurs invisibly, within the first two weeks of pregnancy, long before any imaging technology could observe it directly. What can be observed and studied, once a chimeric individual is identified through whatever combination of visible signs or genetic testing led to their diagnosis, is the downstream pattern of cell distribution that resulted from that early, unobservable fusion event.

Heterochromia, the condition in which a person's two eyes display distinctly different colors, deserves particular attention here, both because it is one of the most frequently cited potential external markers of chimerism in popular discussions of the condition, and because it illustrates nicely how careful researchers and clinicians need to be before attributing any specific physical trait to chimerism without further confirmation. The vast majority of people with heterochromia do not have it because they are chimeras. Complete heterochromia, in which one iris is an entirely different color from the other, most commonly traces back to ordinary inherited genetic variation, to an eye injury sustained after birth, to certain medications, or to a handful of specific medical syndromes entirely unrelated to chimerism, such as Waardenburg syndrome, a genetic condition affecting pigmentation more broadly. Only in a genuine minority of cases does heterochromia trace back to chimerism specifically, and confirming that connection for any individual case requires actual genetic testing rather than visual inspection alone, since heterochromia caused by chimerism looks, to the naked eye, indistinguishable from heterochromia caused by any of its far more common alternative explanations.

When heterochromia does trace back to chimerism, the explanation is fairly intuitive given everything already established about how the condition works. Eye color is determined by the concentration and distribution of a pigment called melanin within the iris, and the genetic instructions governing melanin production can differ between a chimeric individual's two distinct cell lineages, exactly as those same genetic differences can produce Taylor Muhl's two toned skin or the divided fur coloration documented in chimeric cats. If one of a chimeric individual's two cell lineages happens to predominate in the tissue forming one eye, while the other lineage predominates in the tissue forming the second eye, purely as a consequence of how the original fused embryo's cells happened to migrate and differentiate during development, the result can be two eyes that are, genetically speaking, no more similar to each other than the eyes of two ordinary siblings would be, simply housed within a single face rather than two separate ones.

Blood type presents a similarly instructive, and occasionally more clinically consequential, external marker. Some documented chimeric individuals have been found, through routine blood typing procedures conducted for entirely unrelated medical reasons, to carry two distinct populations of red blood cells simultaneously, a condition that shows up in laboratory testing as what appears to be a mixed sample rather than a single, internally consistent blood type. Case reports describing this phenomenon have documented specific ratios of the two blood types coexisting within a single chimeric individual's bloodstream, with one well documented case showing a majority of type O blood cells coexisting alongside a smaller but still substantial minority population of type A cells, a pattern that would be genuinely impossible in a non chimeric individual, whose blood type, whatever it happens to be, remains uniform throughout their entire body precisely because every blood cell they produce descends from the same single genetic lineage. For a chimeric individual, this kind of mixed blood type finding carries genuine practical significance well beyond scientific curiosity, since it directly affects how blood transfusions and organ transplant compatibility testing need to be interpreted and managed, a point already illustrated vividly by Karen Keegan's kidney transplant compatibility crisis in an earlier chapter of this book.

Beyond heterochromia and mixed blood type, researchers cataloguing the range of physical and physiological signs associated with confirmed chimerism cases have documented an unusually broad and varied list of additional possible markers, no single one of which is common enough or specific enough to serve as reliable proof of chimerism on its own, but which collectively illustrate just how many different ways the condition can manifest depending on the particular pattern of cell distribution involved in any given case. These documented markers include patches of hair on different parts of the scalp or body displaying distinctly different natural textures or colors from the surrounding hair, mismatched or unusual patterns of skin pigmentation considerably less dramatic and less cleanly defined than Taylor Muhl's midline division, subtle asymmetries in body or limb size between the left and right sides of an affected individual, and, in a small number of especially striking documented cases, the presence of ambiguous or mixed reproductive anatomy, arising in situations where the two original fused embryos happened to be of different chromosomal sexes, one carrying a typical female chromosome pattern and the other a typical male pattern, a circumstance that can produce combined or intermediate reproductive characteristics depending on how the two genetically distinct cell lineages happen to distribute across the developing reproductive organs.

This same category of sex chimerism, arising when a female embryo and a male embryo fuse into a single developing individual, has been documented in a small number of published human case reports, generally identified through investigation of otherwise unexplained physical characteristics or through a mismatch discovered between an individual's chromosomal sex, as determined through genetic testing, and outward physical appearance. Researchers examining such cases have noted that the resulting chimeric individual will typically undergo a natural biological process similar in principle to the X inactivation process described in the context of chimeric cats earlier in this book, in which competing genetic signals from the two distinct cell lineages must somehow resolve into a single, functioning, and generally outwardly consistent physical body, a process that can produce a range of different outcomes depending on the precise proportions and locations of each lineage's contribution to the developing embryo's reproductive and other tissues.

Given how genuinely varied these potential physical markers are, and given that a meaningful number of confirmed chimeric individuals display no detectable external sign whatsoever, researchers and genetic counselors who specialize in chimerism consistently emphasize a note of caution directly relevant to any reader of this book who might, after learning about these various markers, begin wondering whether some minor physical quirk of their own could indicate undiagnosed chimerism. The overwhelming majority of heterochromia cases, skin pigmentation variations, hair texture differences, and body asymmetries encountered in the general population have entirely ordinary, non chimeric explanations, and no physical trait discussed in this chapter should be treated as anything resembling a reliable, standalone diagnostic signal. Confirming actual chimerism requires genetic testing comparing multiple distinct tissue samples from the same individual, precisely the kind of testing that, as later chapters of this book will explore in more detail, most people never have occasion to undergo unless some specific triggering circumstance, whether medical, legal, or simply personal curiosity following a period of unexplained physical distinctiveness, prompts it.

Blood chimerism deserves particular mention here, because it represents one of the more medically consequential physical manifestations the condition can produce. Some chimeric individuals carry two distinct populations of red blood cells simultaneously, a condition detectable through blood typing tests that reveal what appears to be a mixed sample rather than a single consistent blood type. This can create genuine complications in medical contexts where blood type matters significantly, including blood transfusion compatibility testing and, as demonstrated vividly in Karen Keegan's case, organ transplant compatibility screening. A chimeric individual undergoing a blood transfusion, or serving as a potential organ donor or recipient, may present a genuinely confusing profile to medical staff unfamiliar with the condition, underscoring the practical, real world stakes of expanding medical awareness of chimerism beyond a small community of genetics specialists.

Taylor Muhl's public visibility has also helped illuminate just how significant the gap likely is between documented chimerism cases and the true, largely unmeasured prevalence of the condition in the general population, a theme introduced in earlier chapters and worth revisiting here with her case specifically in mind. Taylor's chimerism was, in a meaningful sense, hiding in plain sight for the majority of her life. It produced a visible physical sign present from birth, examined by multiple physicians over the years, and still took decades to correctly identify, purely because chimerism was not part of the standard diagnostic vocabulary most of those physicians carried with them into the examination room. If a condition producing a visible, birth present physical marker can go misdiagnosed as an ordinary birthmark for decades, it raises an obvious and unsettling question about how many chimeric individuals exist whose condition produces no visible sign whatsoever, and who therefore have essentially no pathway toward diagnosis at all, absent some unrelated event, like a legal dispute or a failed organ compatibility test, that happens to trigger genetic testing of the relevant tissue.

Taylor has spoken about wanting her visibility and advocacy to help close exactly this gap, encouraging both the public and medical professionals to consider chimerism as a possible explanation when confronted with otherwise inexplicable physical asymmetries, immune irregularities, or genetic testing anomalies. Her case stands as one of the clearest illustrations available of what chimerism can look like when it does produce visible external signs, and as a reminder that even visible signs are no guarantee of timely or accurate diagnosis in a medical system generally unfamiliar with a condition this rare.

The chapters that follow move from these dramatic, fused twin forms of chimerism toward related but mechanistically distinct phenomena: blood chimerism between twins who remain separate individuals rather than fusing, and the universal, far more common form of chimerism that occurs during pregnancy itself, in which mother and child exchange cells regardless of whether either was ever part of a twin pregnancy at all.

CHAPTER 6

Blood, Marrow, and Borrowed Cells

A pregnant woman in her late thirties, conceiving through intracytoplasmic sperm injection after two good quality embryos were transferred into her uterus, went in for a routine ultrasound at six weeks and received the news that she was carrying twins sharing a single placenta. Doctors classified the pregnancy as monochorionic, meaning both twins depended on the same placental structure for nourishment, a classification that typically, though not always correctly, is assumed to indicate identical twins descended from a single fertilized egg that split in two. As the pregnancy progressed, a later ultrasound at twenty five weeks revealed something that should not have been possible under that assumption: the two twins appeared to be of different sexes.

Identical twins share the same genetic sex, because they originate from the same single fertilized egg. Twins of different sexes are, by definition, fraternal, arising from two separately fertilized eggs. And yet here were two twins sharing one placenta, a physical arrangement overwhelmingly associated with identical twinning, who were nonetheless clearly fraternal based on their differing sexes observed directly on ultrasound.

This case, documented in the medical literature and cited by researchers studying rare pregnancy complications, illustrates a category of chimerism distinct from the tetragametic fusion described in earlier chapters, though related to it in important ways. Following delivery, genetic testing of the twins' blood revealed that both infants carried a mixture of two distinct karyotypes, one with a female chromosome pattern and one with a male chromosome pattern, present simultaneously in each twin's bloodstream. Blood chimerism had occurred between two twins who otherwise remained entirely separate individuals, each with their own body, their own developmental trajectory, and their own distinct genetic identity in every tissue other than blood.

Notably, when researchers subsequently tested cells from the inside of each twin's cheek, using a simple buccal swab, no chimerism was detected at all. Each twin's cheek cells showed only their own individual genetic profile, with no trace of their sibling's DNA. The chimerism in this case was confined entirely to the bloodstream, a distinction that matters enormously for understanding how this form of chimerism differs from the tetragametic fusion chimerism examined in the opening chapters of this book.

The explanation lies in a well documented, if still only partially understood, pregnancy complication called twin to twin transfusion syndrome. This condition arises specifically in monochorionic twin pregnancies, where two developing fetuses share a single placenta and, critically, share a network of blood vessels running through that placenta that connect their two separate circulatory systems. In an ordinary monochorionic pregnancy, blood flows across these connecting vessels in a roughly balanced fashion, and both twins receive adequate nourishment without significant complication. In twin to twin transfusion syndrome, the blood flow across these shared vessels becomes imbalanced, with one twin, termed the donor, losing a disproportionate share of blood volume to the other twin, termed the recipient. The donor twin becomes volume depleted, while the recipient twin becomes volume overloaded, a serious complication that obstetric specialists estimate occurs in roughly ten percent of monochorionic pregnancies and that requires careful monitoring and, in more severe cases, active medical intervention such as fetoscopic laser surgery to seal off the problematic connecting vessels before the imbalance causes serious harm to either twin.

Twin to twin transfusion syndrome is overwhelmingly discussed in the medical literature, and in this book's earlier mention of the condition, as a complication of identical, monozygotic twin pregnancies, since the vast majority of monochorionic pregnancies do involve identical twins who split from a single fertilized egg early enough in development to end up sharing one placenta while remaining in two separate amniotic sacs. But the case described above, along with several other documented cases in the specialized medical literature on rare twin pregnancy complications, demonstrates that monochorionic placentation can occasionally occur even in dizygotic, fraternal twin pregnancies, through mechanisms that remain incompletely understood but that appear to be more likely in pregnancies conceived through assisted reproductive technology, particularly procedures involving assisted hatching of the embryo before transfer into the uterus.

When twin to twin transfusion syndrome occurs in a monochorionic dizygotic pregnancy, meaning two genetically distinct fraternal twins sharing a single placenta, the abnormal blood flow across the shared placental vessels does not simply move plasma and nutrients between the twins. It moves actual blood forming stem cells, the cells responsible for generating all the different types of blood cells throughout a person's life. When these blood forming stem cells cross from one twin into the other through the shared placental circulation, they can, in some cases, successfully engraft in the recipient twin's bone marrow, establishing a permanent, self sustaining population of blood cells derived from a genetically distinct sibling, alongside the recipient's own naturally occurring blood cell population. The result is a form of permanent blood chimerism, in which an individual's blood, and only their blood, contains two genetically distinct cell populations, one belonging to themselves and one belonging to their twin sibling, while the rest of their body, including tissue as accessible as the inside of the cheek, remains entirely and exclusively their own genetic material.

This form of blood chimerism differs meaningfully from the tetragametic chimerism examined in the opening chapters in a crucial respect: the two individuals involved remain two separate people. Unlike Lydia Fairchild or Karen Keegan, whose two genetic lineages coexist within a single body because two originally separate embryos physically fused together before differentiating into distinct tissues, blood chimeric twins are born as two distinct infants, each developing into their own separate person, who simply happen to carry a permanent trace of each other circulating in their blood for the rest of their lives. Documented cases have found this blood chimerism persisting for at least six to nine months after birth in the infants studied, and researchers who study the phenomenon suspect, based on parallel research into other forms of chimerism discussed later in this chapter, that it may persist considerably longer, quite possibly for a lifetime, in at least some cases.

The distinction between blood chimerism confined to one tissue type and full tetragametic chimerism distributed throughout the body has important practical implications. A blood chimeric individual, tested via a standard blood draw, might show a confusing mixed genetic signature exactly like Lydia Fairchild's or Karen Keegan's blood did. But if that same individual's cheek cells, hair, or reproductive tissue were tested instead, the result would show only their own unambiguous genetic profile, with no trace of a twin's DNA whatsoever, because the chimerism in these cases never spread beyond the blood forming system in the first place. This means that blood chimerism resulting from twin to twin transfusion syndrome, while genuinely capable of producing exactly the kind of confusing DNA test result that could trigger a legal or medical crisis similar to Lydia Fairchild's, is mechanistically distinct from, and considerably less pervasive throughout the body than, tetragametic chimerism resulting from early embryonic fusion.

The broader category of chimerism arising from shared placental blood supply extends beyond twin to twin transfusion syndrome specifically. Any monochorionic twin pregnancy involves some degree of blood vessel connection between the two developing fetuses, and even in pregnancies that never develop the severe blood flow imbalance that characterizes clinically diagnosed twin to twin transfusion syndrome, some low level exchange of blood cells between twins sharing a placenta appears to be common, and in cases involving genetically identical monozygotic twins, this exchange produces no detectable chimerism at all, since both twins already carry identical DNA and there is no genetic mismatch to observe even if cells do cross between them. It is specifically in the rarer cases of dizygotic twins who nonetheless end up sharing a monochorionic placenta, through whatever developmental quirk produces that unusual arrangement, that blood chimerism becomes both detectable and, from a research perspective, genuinely informative about how thoroughly placental blood vessel connections can merge two developing immune and blood systems.

Cattle and other livestock species have long provided the clearest and most extensively studied model of this phenomenon, because in cattle, fraternal twin pregnancies routinely and normally develop shared, fused placental blood vessel connections, a pattern that occurs far more commonly and predictably in cattle than in humans. Veterinary researchers documented blood chimerism in cattle twins many decades before the phenomenon was well characterized in humans, and this cattle research, particularly involving cases where a female calf twinned with a male calf, produced a well known related phenomenon called freemartinism, in which hormonal and cellular exchange between the male and female twin fetuses, carried across the shared placental blood supply, disrupts normal reproductive development in the female twin. The extensive veterinary literature on cattle twin pregnancies, developed initially for entirely practical agricultural reasons related to livestock breeding, ended up providing much of the foundational biological understanding that researchers later applied to interpreting the rarer and more sporadically documented cases of blood chimerism found in human twin pregnancies.

The freemartin phenomenon in cattle deserves a somewhat closer look, both because of its historical importance to the broader scientific understanding of chimerism and because it illustrates, in a species where the underlying biological mechanism occurs reliably and can therefore be studied systematically rather than only encountered incidentally, just how profoundly shared placental blood exchange between twins can affect development beyond the blood system alone. A freemartin is a female calf born as a twin to a male calf sharing the same placenta, and freemartin calves are, in the overwhelming majority of documented cases, reproductively infertile, owing to hormones and cellular signals transmitted from the male twin across the shared placental blood supply that interfere with normal development of the female twin's reproductive organs during a critical developmental window. This phenomenon, well recognized by cattle breeders long before its underlying genetic and hormonal mechanism was scientifically understood, provided some of the earliest practical, real world evidence that twins sharing a placenta could meaningfully affect one another's development beyond simple nutrient competition, a foundational insight that later informed how researchers came to understand and investigate the more subtle and less visually dramatic forms of blood chimerism documented in human twin pregnancies.

Twin pregnancies involving a shared placenta also carry a distinct set of monitoring requirements and delivery considerations quite different from those involved in ordinary, separately implanted twin pregnancies, a difference worth briefly noting here given how directly it connects to the specific mechanism producing blood chimerism in these pregnancies. Obstetricians managing a monochorionic twin pregnancy typically recommend considerably more frequent ultrasound monitoring throughout the second and third trimesters than would be standard for a dichorionic pregnancy, in which each twin has its own separate placenta, precisely because the shared blood vessel network connecting monochorionic twins creates ongoing risk not just for the twin to twin transfusion syndrome discussed throughout this chapter, but for a range of related complications that can develop with relatively little warning between scheduled appointments. This heightened monitoring standard reflects decades of accumulated clinical experience demonstrating that monochorionic twin pregnancies, while representing only a minority of all twin pregnancies, account for a disproportionate share of serious twin pregnancy complications, making them a recognized higher risk category deserving of correspondingly more intensive obstetric attention throughout gestation.

What does living with blood chimerism actually mean for someone who carries it, assuming it persists into adulthood as researchers suspect it often does? The honest answer, based on the current state of medical research, is that in most documented cases, it appears to mean remarkably little in terms of day to day health or functioning. Blood chimerism resulting from twin to twin transfusion syndrome does not, based on available evidence, produce the kind of immune confusion or graft versus host complications that might be expected when foreign blood cells are deliberately introduced into an adult's body through a mismatched transfusion or transplant. This is likely because the exchange occurs so early in fetal development, before either twin's immune system has fully matured and established its definitive sense of self versus foreign material, that both the individual's own blood cells and their twin's blood cells circulating within them are recognized by the developing immune system as legitimately belonging to the body, a form of immune tolerance established through exposure during a critical developmental window rather than after the fact.

This same principle of early developmental immune tolerance, in which cells introduced before the immune system fully matures come to be treated as self rather than foreign, turns out to be central to understanding an even more widespread and universally relevant form of chimerism than anything discussed so far in this book: the microchimerism that occurs during essentially every human pregnancy, regardless of whether twins are involved at all, in which mother and developing child exchange cells that persist, quietly and almost universally undetected, for years or decades after birth. That phenomenon, touching a far larger population than the rare twin based forms of chimerism examined in this chapter, is the subject explored next.

Before turning to that broader phenomenon, it is worth briefly considering how blood chimerism resulting from twin to twin transfusion syndrome is typically identified and managed in modern clinical practice, since the answer illustrates a genuinely encouraging development within an otherwise often frustrating diagnostic landscape. Unlike tetragametic fusion chimerism, which as earlier chapters have discussed tends to be discovered incidentally, often years or decades after birth, blood chimerism arising from twin to twin transfusion syndrome is considerably more likely to be identified relatively soon after birth, precisely because monochorionic twin pregnancies already receive intensive prenatal monitoring specifically due to the well recognized risk of twin to twin transfusion syndrome itself. Obstetricians managing a monochorionic twin pregnancy routinely perform frequent ultrasound examinations throughout the second and third trimesters specifically to monitor for signs of blood flow imbalance between the twins, meaning that even in the rarer dizygotic monochorionic cases like the one that opened this chapter, the medical team is already primed to expect and investigate exactly the kind of complication that produces detectable blood chimerism, in sharp contrast to the complete absence of any comparable monitoring protocol for tetragametic fusion chimerism, which offers no equivalent early warning signal during pregnancy at all.

This difference in typical detection timing carries a meaningful silver lining for families affected by dizygotic monochorionic twin pregnancies specifically. Because these pregnancies already receive heightened medical scrutiny, any blood chimerism that develops is considerably more likely to be identified relatively early in a child's life, whether through the kind of genetic sex mismatch that alerted researchers to the case described at the opening of this chapter, or through more routine newborn blood typing that happens to reveal an unexpected mixed profile. This stands in useful contrast to the pattern documented repeatedly throughout the rest of this book, in which tetragametic fusion chimerism typically surfaces only by accident, often decades after birth, through some unrelated legal or medical crisis that happens to trigger the right kind of genetic testing at the right time.

CHAPTER 7

Every Mother Carries Her Children

Researchers examining the blood of women decades past childbirth have found something that quietly rewrites the boundary of where one person ends and another begins: cells belonging to their sons, still circulating, still alive, still functionally active, found in mothers whose youngest child had turned twenty seven years old.

This finding, reported by researchers studying a phenomenon called fetal microchimerism, was not an isolated fluke. It reflects a now well established and extensively replicated body of research showing that during pregnancy, cells cross the placental barrier in both directions, from mother to fetus and from fetus to mother, and that a meaningful population of these exchanged cells does not simply pass through and disappear but takes up permanent, long term residence in the recipient's body, continuing to function and, in some cases, to divide and multiply for years or decades after the pregnancy that introduced them has ended.

This is chimerism, in the fullest technical sense of the term, occurring not as a rare curiosity affecting a documented handful of individuals worldwide, but as a nearly universal feature of human pregnancy, touching, at some measurable level, every mother who has ever carried a child and quite possibly every child who has ever been carried.

The mechanism differs fundamentally from the tetragametic fusion chimerism examined in earlier chapters. Microchimerism does not require two embryos merging into one. It requires only an ordinary, singleton pregnancy, and it exploits the simple physical fact that the placenta, while an effective barrier against many pathogens and large molecules, is not a perfectly sealed boundary between the fetal and maternal bloodstreams. Throughout gestation, and increasingly as pregnancy progresses toward term, small numbers of cells cross this boundary in both directions. Researchers have found that a measurable percentage of the cell free DNA floating in a pregnant woman's blood plasma, a percentage that can reach into the mid single digits by some estimates, originates from the developing fetus rather than from the mother herself, a finding so robust and clinically useful that it now forms the biological basis for noninvasive prenatal genetic testing offered to pregnant women worldwide, allowing doctors to screen for certain chromosomal conditions in a developing fetus using nothing more than a blood sample from the mother.

Most of the fetal cells that enter the maternal bloodstream during pregnancy do not persist long term. The majority are cleared by the mother's immune system, through a normal process of cell death called apoptosis, within a relatively short window after they cross the placental barrier. But a subset of these cells, for reasons researchers are still working to fully characterize, evade this clearance process and establish lasting residence within specific maternal tissues, effectively becoming permanent, functioning residents of organs throughout the mother's body.

The list of maternal organs in which researchers have identified persistent fetal cells is extensive and, for many people encountering this research for the first time, genuinely startling: the liver, the kidney, the spleen, the lung, the heart, the skin, and perhaps most strikingly, the brain. Researchers using techniques capable of detecting the presence of a Y chromosome, which provides an unambiguous genetic marker in women who have given birth to sons, since the mother herself carries no Y chromosome of her own, have identified male fetal cells embedded within multiple distinct regions of the maternal brain, in at least one documented case detected decades after the woman had given birth to the son whose cells were found there. This finding led researchers studying the phenomenon to suggest that fetal microchimerism within the human maternal brain may be considerably more pervasive and more durable than earlier, more limited studies had suggested.

This particular finding, fetal cells persisting for decades within specific regions of the maternal brain, deserves to be sat with a little longer than a purely clinical summary might suggest, given how directly it touches on some of the deeper philosophical themes this book returns to in its later chapters. A mother's own neural tissue, the physical substrate of her memories, her thoughts, and her sense of herself, can, according to this research, contain living cells that originated not from her own original zygote, but from a pregnancy she carried years or decades earlier. This does not mean, and researchers studying the phenomenon have been careful never to suggest, that a mother's thoughts or personality are in any sense altered or shared with her child through this cellular presence. The functional significance of these persistent neural fetal cells, whether they contribute anything at all to ongoing brain function or simply exist quietly alongside the mother's own original neurons without exerting any detectable influence, remains a genuinely open scientific question that current research has not yet definitively answered. What the finding does establish, with reasonable confidence, is that the physical, cellular boundary between a mother's brain and her child's genetic contribution to that brain is less absolute than most people, including many with a general familiarity with pregnancy and genetics, would likely assume before encountering this specific area of research.

The reverse process, maternal microchimerism, in which the mother's own cells cross the placenta into the developing fetus and establish similarly persistent residence within the child's body, occurs alongside fetal microchimerism, though research suggests the exchange is asymmetrical, with a greater volume of cells typically moving from fetus to mother than from mother to fetus. Maternal cells that establish themselves within a developing child's tissues can persist, according to current research, for decades after birth as well, meaning that essentially every person born through a natural pregnancy likely carries, throughout their life, a small but real and measurable population of cells that originated not from their own body's original genetic material, but from their mother's.

This means that the boundary most people intuitively imagine between a parent and a child, one body ending and a separate body beginning at the moment of birth, or even at the moment of conception, is considerably less clean than everyday experience suggests. At a cellular level, mother and child continue exchanging and permanently retaining small populations of each other's living tissue for years, and by some evidence decades, after the pregnancy that connected them has concluded. Multiple generations within a single family, in principle, could carry detectable traces of each other's cells passed forward and backward across successive pregnancies, a possibility that researchers studying microchimerism have begun to investigate more directly as detection techniques improve.

What, if anything, do these persistent foreign cells actually do once they have taken up residence in a parent's or child's body? This question sits at the frontier of current microchimerism research, and the honest scientific answer involves genuine uncertainty alongside a growing body of intriguing, if not yet fully conclusive, findings.

Some research suggests that microchimeric cells play a beneficial role in maternal health, potentially participating directly in tissue repair. Researchers studying heart tissue, for example, have investigated whether fetal cells that migrate into a mother's heart during pregnancy might contribute to repairing damage from cardiac injury, effectively acting as an unexpected reservoir of regenerative cells positioned within maternal tissue as a byproduct of pregnancy itself. Other research has focused on the immunological role microchimeric cells may play, with findings suggesting that maternal cells persisting within offspring may help establish a form of immune tolerance, teaching the child's developing immune system not to attack antigens that are present in the mother but that the child did not directly inherit, a mechanism that researchers studying this effect suggest may play a protective role in future pregnancies carried by female offspring later in their own lives, helping their bodies better tolerate a fetus carrying paternal genetic material that the mother's own immune system might otherwise be more inclined to recognize as foreign.

Alongside this evidence of potentially beneficial roles, researchers have also documented associations between microchimerism and certain autoimmune conditions, most notably scleroderma, a disease involving hardening and tightening of the skin and connective tissue that shares some clinical features with graft versus host disease, the serious complication that can occur when transplanted tissue attacks a recipient's body. Early research identified persistent maternal microchimerism at notably elevated rates among patients diagnosed with scleroderma compared to the general population, leading researchers to hypothesize that in at least some cases, long dormant maternal cells persisting in an adult child's body might, under conditions not yet fully understood, become immunologically active in a way that contributes to autoimmune tissue damage resembling a low grade, chronic version of transplant rejection occurring entirely within a person's own body, using cells that have resided there since before birth. This hypothesis remains an active area of investigation rather than settled scientific consensus, and researchers are careful to note that the relationship between microchimerism and autoimmune disease appears complex and almost certainly bidirectional in its associations, meaning some findings suggest microchimeric cells may worsen certain conditions while other findings suggest they may play protective roles depending on the specific tissue, disease, and individual circumstances involved.

Researchers have also explored a striking, and admittedly more speculative, line of investigation into whether fetal microchimerism might play a meaningful role in maternal cancer risk and outcomes, particularly in breast cancer. Some studies have found that fetal cells appear at meaningfully elevated concentrations specifically within breast tumor tissue compared to the surrounding healthy breast tissue in the same patient, a finding that researchers have interpreted in two quite different, competing ways. One interpretation suggests fetal cells might be recruited to tumor sites as part of a natural tissue repair or immune surveillance response, potentially playing a protective role against cancer progression. A competing interpretation suggests the opposite: that the altered immune environment surrounding a tumor might simply provide more favorable conditions for fetal cells to survive and proliferate, making their presence a consequence rather than a cause of the tumor's development, with no protective function at all. This remains a genuinely open scientific question, and readers should treat popular claims in either direction, including some found in less rigorous sources, with appropriate scientific caution until more definitive research clarifies the relationship.

Beyond disease research, microchimerism carries genuinely poignant implications that extend past clinical medicine into something closer to the philosophical territory this book will explore more directly in later chapters. A woman who has miscarried, or who has terminated a pregnancy, may nonetheless carry persistent fetal cells from that pregnancy for years afterward, a finding that researchers studying pregnancy loss have noted carries real emotional resonance for many women who learn of it, offering an unexpected form of continued biological connection to a pregnancy that did not result in a living child. Similarly, a mother who has raised several children may carry, simultaneously, small persistent populations of cells from multiple different pregnancies, meaning that in a very literal, measurable, cellular sense, she carries a physical trace of each of her children within her own body for the remainder of her own life, layered chronologically according to when each pregnancy occurred.

Researchers studying the emotional and psychological dimensions of microchimerism, an area of inquiry considerably newer and less developed than the underlying cell biology research itself, have begun exploring how awareness of this phenomenon affects people who learn about it in various circumstances, whether through direct participation in a research study, through a healthcare provider mentioning it in the context of an unrelated medical consultation, or simply through general science reporting of the kind this book aims to provide. Early qualitative research in this area, though still limited in scope, has suggested that many people react to learning about microchimerism with a mixture of scientific fascination and genuine emotional resonance, particularly parents reflecting on their own relationships with their children, and adult children reflecting on their relationship with a mother who may have since passed away, since microchimeric cells inherited during gestation persist as a literal, ongoing biological connection extending beyond a parent's death in a way that few other forms of inheritance can claim to do quite so directly or so tangibly.

Detection of microchimerism has historically relied heavily on identifying male, Y chromosome carrying cells within female tissue, since this provides an unambiguous genetic marker requiring no advance knowledge of the specific individual's genetic profile beyond their sex. This methodological reliance has created a meaningful gap in the research: microchimerism resulting from pregnancies involving daughters is considerably harder to detect using standard techniques, since both mother and daughter share the same chromosomal sex, requiring far more sophisticated genetic techniques capable of distinguishing subtler differences between individual genetic profiles that do not involve something as clearly binary as the presence or absence of a Y chromosome. Some research has found evidence suggesting that microchimeric cell quantities may actually be higher following pregnancies involving female fetuses compared to male fetuses, a finding that, if confirmed through further research using improved detection methods, would suggest that decades of microchimerism research relying primarily on Y chromosome detection may have meaningfully underestimated the true scope and prevalence of the phenomenon across the general population of mothers and children.

The relationship between a woman's complete pregnancy history and the total population of microchimeric cells she carries also deserves attention, because it complicates any simple picture of microchimerism as a straightforward, one to one exchange between a single mother and a single child. Researchers have found evidence suggesting that a woman's body can retain detectable microchimeric cell populations from multiple separate pregnancies simultaneously, layered chronologically according to when each pregnancy occurred, meaning a mother of several children may carry, at any given point in her adult life, a composite population of cells drawn from more than one of her children at once, each population persisting at its own pace and its own gradually declining concentration over time. Some research has extended this line of inquiry even further, raising the genuinely striking possibility that a woman might carry detectable microchimeric cells originating not from her own pregnancies at all, but transmitted to her originally from her own mother during the pregnancy that produced her, effectively passed down across more than one generation, a phenomenon researchers have described as grandmaternal microchimerism. Confirming this multigenerational transmission with full scientific rigor remains genuinely challenging, since distinguishing a grandmother's cells specifically from a mother's own cells, or from a sibling's cells retained from an earlier pregnancy, requires unusually precise and carefully controlled genetic analysis, but the research conducted so far has been sufficiently suggestive that it remains an active and genuinely intriguing area of ongoing investigation.

Pregnancy itself, beyond simply being the mechanism through which microchimerism initially arises, appears also to influence how a woman's own body responds to and manages microchimeric cells retained from earlier pregnancies, according to research examining how microchimeric cell populations shift and change across successive pregnancies within the same woman. Some researchers have investigated whether cells retained from an earlier pregnancy might play a functional role in a woman's body during a subsequent, later pregnancy, findings that connect back to the immune tolerance research discussed earlier in this chapter, since a woman's immune system managing a current pregnancy is simultaneously managing whatever residual microchimeric cell populations remain from her own mother's pregnancy with her, and potentially from any of her own earlier pregnancies as well, layering an already complex immunological balancing act with additional genetic populations that most textbook descriptions of pregnancy immunology, focused narrowly on the immediate mother and current fetus, do not typically account for in their standard explanatory framework.

Advances in detection technology continue to expand what researchers can observe. Techniques capable of identifying and counting individual microchimeric cells throughout an entire developing embryo, rather than sampling limited tissue sections, have recently allowed researchers to observe striking individual variation in how many maternal cells different embryos from the same pregnancy or from different pregnancies carry, with at least one documented case showing a single embryo carrying several hundred times more maternal cells than typical, for reasons not yet understood. This kind of finding underscores how much about microchimerism remains genuinely unexplored, despite the phenomenon itself having been observed, in some early form, since researchers first identified fetal cells in the lungs of women who had died from a serious pregnancy complication called eclampsia, in research dating back to the final years of the nineteenth century, more than a century before modern genetic techniques allowed the phenomenon to be studied with anything approaching its current level of precision.

What emerges from this body of research is a picture of human pregnancy as something considerably more biologically intimate and lasting than the nine months typically associated with gestation. Every pregnancy that proceeds far enough for placental blood exchange to occur, which is to say essentially every pregnancy that continues past the earliest weeks, appears to leave behind a permanent cellular trace in both directions, mother in child and child in mother, persisting quietly for years and, in a growing number of documented cases, for decades. This is chimerism at its most universal and its most easily overlooked, a biological reality so common that virtually no person who has been born through pregnancy, or who has carried a pregnancy themselves, entirely escapes it, and yet so invisible in ordinary life that almost nobody thinks to consider it outside of specialized medical research.

The next chapter turns from the biology of chimerism toward its collision with human institutions, examining how legal systems, built around assumptions of genetic uniformity that chimerism directly violates, have struggled to accommodate a phenomenon that DNA testing, for all its scientific power, was never originally designed to detect or explain.

CHAPTER 8

When DNA Lies to the Law

Police in Alaska had, by every measure that normally settles a criminal case, an airtight match. Semen recovered from a 2005 sexual assault investigation had been run against a state DNA database, and it returned a hit. The genetic profile matched a man already in the system. There was just one problem, and it was not a small one. At the time of the assault, the man the DNA identified was sitting in jail.

An alibi this solid should have ended the investigation immediately, with detectives concluding that the database match was either a coincidence involving an unrelated individual with a similar profile, an outcome statistically almost impossible given how DNA profiling works, or an error somewhere in the chain of evidence handling. Neither explanation turned out to be correct. Investigators eventually determined that the incarcerated man had, years earlier, donated bone marrow. His actual biological brother, it emerged, had committed the assault, and the brother's semen carried a genetic profile that matched the incarcerated donor rather than the brother's own original DNA, because a bone marrow transplant had, in an entirely different and much more literal sense than doctors typically intend when discussing successful treatment outcomes, replaced significant portions of the recipient's genetic identity with that of the donor.

This case, documented and discussed in forensic science literature examining the intersection of chimerism and criminal investigation, illustrates a category of chimerism entirely distinct from anything discussed in earlier chapters. It is not congenital, meaning it is not present from birth or built into a person's biology through the fusion of twin embryos or the placental exchange of pregnancy. It is acquired, deliberately introduced through modern medical treatment, and it can be every bit as capable of producing confusing, contradictory, and in this case actively misleading genetic evidence as any naturally occurring form of chimerism examined so far.

This distinction between congenital and acquired chimerism deserves to be stated with particular clarity here, since the remainder of this chapter will move fluidly between both categories in examining their shared legal and forensic consequences. Congenital chimerism, the primary subject of the earlier chapters of this book, originates before birth, whether through the tetragametic fusion of two separately fertilized embryos, through blood exchange between twins sharing a placenta, or through the near universal microchimerism of ordinary pregnancy. Acquired chimerism, by contrast, originates at some point after birth, generally through a deliberate and clinically necessary medical intervention such as bone marrow transplantation, and it can arise in an individual whose body was, prior to that intervention, entirely genetically uniform in the ordinary, non chimeric sense. Both categories share the same essential defining feature, the coexistence of two genetically distinct cell populations within a single body, and both can produce the same category of confusing or misleading forensic and legal consequences, but they differ meaningfully in their underlying cause, their typical age of onset, and, as this chapter will explore, the specific practical steps investigators and legal systems can take to anticipate and correctly interpret each type when it arises.

Bone marrow transplantation, more precisely termed hematopoietic stem cell transplantation, is a medical procedure used to treat a range of serious blood cancers and other blood disorders, including leukemia. The procedure works by destroying a patient's own diseased bone marrow, the tissue responsible for producing blood cells, and replacing it with healthy blood forming stem cells from a donor, whether a close genetic relative or, increasingly common through international donor registries, an unrelated but sufficiently compatible stranger. When the transplant succeeds, the donor's stem cells take up permanent residence in the recipient's bone marrow and begin producing new blood cells indefinitely, effectively replacing the recipient's own blood producing system with a genetically distinct one for the remainder of that patient's life.

The immediate medical consequence is, in the majority of successful cases, a life saved from an otherwise fatal blood cancer. The less immediately obvious but medically well documented consequence is that the transplant recipient becomes, from that point forward, a permanent chimera, carrying their own original genetic material in the majority of their body's tissues while carrying their donor's genetic material specifically within their blood and immune system.

Chris Long, a leukemia patient who received a bone marrow transplant, became one of the more thoroughly documented cases of this phenomenon after subsequent testing revealed the extent to which his donor's DNA had spread through his body. Swabs taken from his lips and cheeks showed a mixture containing both his own original DNA and his donor's DNA. Testing of his semen and blood, however, revealed something more striking still: these samples showed his donor's DNA exclusively, with no detectable trace of his own original genetic profile remaining at all. The blood forming stem cells transplanted into Chris Long's body had not simply supplemented his own blood cell production. They had, for practical purposes, entirely replaced it, to the point where a genetic test of his blood or semen would identify his donor, not him.

The forensic and legal implications of this kind of acquired chimerism extend well beyond the deliberately constructed scenario of the Alaska case. In 2008, authorities in Seoul, South Korea, faced a genuinely bewildering identification puzzle following a car accident. The victim's blood samples indicated a female genetic profile, while tissue recovered from the kidney indicated a male profile. Further testing of spleen and lung tissue revealed a mixture containing both genetic signatures simultaneously. For a period during the investigation, forensic examiners were confronted with evidence that appeared to describe two different people occupying what was unmistakably a single body. The eventual resolution traced back to a bone marrow transplant the victim, a man, had received from his own daughter, whose blood forming stem cells had successfully engrafted and taken over his blood production, leaving his blood chromosomally and genetically female while the rest of his body's tissue, including his kidney, remained his own original, male genetic material.

These cases illustrate why forensic scientists and legal experts who specialize in DNA evidence have increasingly had to grapple with hematopoietic stem cell transplantation as a genuine complicating factor in criminal investigations and legal proceedings that rely on genetic testing. The number of people worldwide who have received such transplants has grown substantially as the medical procedure has become more widely available and more successful over recent decades, meaning the population of acquired chimeras walking around with two distinct genetic profiles distributed across different tissues of their bodies is neither vanishingly small nor likely to shrink in the years ahead. Forensic researchers studying this issue have specifically recommended that investigators handling any case involving a suspect or victim with a known history of bone marrow transplantation give careful consideration to which biological sample type, blood, buccal cells, hair, nail clippings, or skin, is most likely to yield an accurate genetic identification for that specific individual, since the answer can vary considerably from one transplant recipient to another depending on how thoroughly and how far beyond the blood and immune system the donor's cells have spread throughout the body. Research examining nail and skin epithelial cells from transplant patients has found that donor genetic material can, in some cases, be detected in these tissues as well, though generally at lower and more variable levels than in blood, and some research has suggested this level of donor chimerism in skin and nail tissue may increase gradually over the years following a transplant, adding a further layer of complexity for forensic examiners attempting to interpret genetic evidence from a transplant recipient whose sample was collected long after their original procedure.

Beyond the acquired chimerism produced by bone marrow transplantation, the naturally occurring, congenital forms of chimerism examined throughout the earlier chapters of this book carry their own distinct set of legal and forensic implications, most vividly illustrated by Lydia Fairchild's custody battle, which opened this book, but extending into a broader set of legal contexts that specialists in forensic genetics have had to reckon with as awareness of chimerism has grown.

Paternity and maternity testing, the specific legal context that nearly cost Lydia Fairchild her children, remains the area where naturally occurring chimerism most directly and consequentially intersects with the legal system. A documented case from Albania, examined in the forensic genetics literature, describes a paternity dispute in which an alleged father underwent DNA testing to confirm his biological relationship with a young girl. Two earlier tests had produced contradictory and confusing results before a more thorough analysis, conducted using sixteen distinct genetic markers, revealed that his own genetic profile was itself chimeric, showing three, and in some markers four, distinct genetic variants at locations where a non chimeric individual would show only one or two. His medical history subsequently revealed a prior bone marrow transplant, explaining the confusing mixed profile and ultimately allowing the paternity determination to proceed correctly once forensic examiners understood what they were actually looking at.

Sexual assault investigations present a particularly fraught legal context in which chimerism can complicate the pursuit of justice, as the Alaska case involving the incarcerated bone marrow donor demonstrates vividly. Forensic literature examining this intersection also documents a separate and disturbing category of cases involving suspected sexual abuse of unconscious victims, in which biological evidence collected from the victim was compared against national DNA databases in the hope of identifying an unknown assailant, underscoring how much weight the legal system places on DNA database matches functioning as unambiguous, singular identifiers of one specific individual, an assumption that chimerism, whether congenital or acquired through transplantation, directly and demonstrably violates in a meaningful number of documented real world cases.

The broader legal and forensic community's response to this accumulating body of chimerism related case law and research has been to push for greater awareness and more careful protocol among practitioners, rather than any wholesale abandonment of DNA evidence as a forensic tool, which remains, in the overwhelming majority of cases, extraordinarily reliable and free from the kind of complications chimerism introduces. Legal and forensic experts writing on this subject have generally framed chimerism not as a reason to distrust DNA evidence broadly, but as a reminder that even highly reliable scientific tools carry edge cases that trained practitioners need to understand and account for, rather than treating any single genetic test result as an infallible, context free fact incapable of requiring further interpretation.

This distinction matters enormously for how the legal system should ideally respond to chimerism related complications going forward. Lydia Fairchild's case very nearly resulted in a profound miscarriage of justice, not because DNA testing itself was flawed, but because the institutions relying on that testing, a state welfare agency and initially a family court, treated a single test result from a single tissue type as conclusive and unambiguous proof of a biological fact that turned out to be considerably more complicated once investigators looked more broadly across her body's different tissues. The lesson legal and forensic specialists increasingly draw from her case, and from the parallel forensic cases involving acquired chimerism through transplantation, is not that DNA evidence should be trusted less, but that it should be interpreted with appropriate scientific humility, particularly in the rare cases where an individual's documented medical history, family circumstances, or the sheer implausibility of an initial result, as in Lydia's case, suggest that a more thorough biological investigation might be warranted before drawing final legal conclusions.

Genetic genealogy, a rapidly growing field that uses DNA testing combined with genealogical research to identify unknown suspects or unidentified remains, represents a newer forensic application in which chimerism specifically resulting from bone marrow transplantation has begun receiving focused research attention, given how central accurate genetic profiling is to the entire methodology underlying this investigative approach. Researchers examining this intersection have specifically studied how bone marrow transplantation affects not just the older style of genetic marker traditionally used in forensic DNA profiling, but also the newer, more comprehensive genetic markers increasingly used in genetic genealogy work, finding that transplant recipients can produce genetic profiles significantly complicated enough to require specialized interpretation before genetic genealogy techniques can be reliably applied to their samples.

This growing body of forensic research into acquired chimerism has practical implications reaching well beyond the specific transplant recipients whose cases have already been documented. As genetic genealogy techniques continue to be applied retroactively to unsolved cold cases, some involving biological evidence collected years or even decades before genetic genealogy existed as an investigative technique, researchers have raised the possibility that older evidence samples, collected without any awareness of a suspect's or victim's transplant history, might already exist within evidence archives worldwide, awaiting analysis using a technique that current best practices would recommend interpreting only with knowledge of that medical history in hand. Cold case investigators reopening decades old files using modern genetic genealogy techniques have increasingly incorporated deliberate outreach to medical records, family members, and other sources capable of establishing a subject's transplant history specifically to guard against this kind of retroactive complication, reflecting how thoroughly chimerism awareness has begun to reshape standard investigative practice in this particular and rapidly evolving corner of forensic science.

As both bone marrow transplantation and increasingly sophisticated genetic testing techniques continue to become more widespread and more central to both medicine and criminal justice, the population of people living with some form of chimerism, whether congenital and present since before birth or acquired through medical treatment later in life, will almost certainly continue to grow, and with it, the forensic and legal community's need to maintain careful awareness of a biological phenomenon that, while still statistically rare on a population wide basis, carries outsized consequences whenever it intersects with a legal system built on the underlying assumption that every person carries one, and only one, genetic identity throughout their entire body.

Immigration proceedings represent another significant legal context in which chimerism related complications have surfaced, one that carries particularly high stakes given how heavily many immigration systems worldwide rely on DNA testing to verify claimed family relationships between individuals seeking to reunite with relatives already living in a host country. Family reunification cases often require an applicant to demonstrate a genetic relationship to a sponsoring relative, and the same fundamental vulnerability that nearly cost Lydia Fairchild her children, a single tissue sample yielding a genetic profile that fails to reflect an individual's true, more complex biological reality, carries an equally serious risk within immigration contexts, where a false negative result could permanently separate genuinely related family members or, conversely, a chimerism related complication might in rare cases be exploited to support a fraudulent claim of relationship where none actually exists. Forensic geneticists writing on this subject have specifically flagged immigration DNA testing as an area deserving of the same kind of protocol development and professional awareness building already underway within criminal forensic science, given the significant human consequences that can follow from a mistaken determination in either direction.

Civil litigation involving inheritance and estate disputes offers a further legal context worth mentioning, since these cases frequently hinge on establishing biological relationships, whether between a claimed child and a deceased parent, or between potential heirs and a shared ancestor, using DNA evidence gathered sometimes years or decades after the relationships in question were originally formed. A chimeric individual involved in such a dispute, whether as the person whose relationship to an estate is being questioned or as a family member whose own DNA is being used as a comparison point, could in principle introduce exactly the kind of confusing, contradictory genetic evidence that has already surfaced in the paternity, maternity, and forensic criminal contexts examined throughout this chapter, though documented inheritance cases specifically involving confirmed chimerism remain considerably rarer in the published legal and forensic literature than the paternity and criminal cases that have received more direct scholarly attention to date.

The following chapter shifts from human chimerism toward the deliberately engineered chimeras that scientists have created in laboratories for decades, a body of research that has produced some of modern biology's most important tools and raised some of its most persistent ethical questions.

CHAPTER 9
The Made Chimeras

Under ultraviolet light, part of the pancreas glowed green.

The organ belonged to a rat, but roughly all of it had actually grown from mouse cells, engineered to fluoresce so researchers could track exactly which cells built which structures. Scientists had taken a rat embryo genetically modified so that it could not grow a pancreas of its own, then injected mouse embryonic stem cells, tagged with a green fluorescent protein originally derived from jellyfish, directly into that embryo at an early developmental stage. The rat embryo, still capable of everything else required to build a normal rat body, filled the pancreatic gap in its own development using the injected mouse cells instead, producing an adult rat with a fully functional, fully rat sized pancreas built entirely from mouse derived tissue. When researchers later transplanted that glowing, mouse grown pancreas into a diabetic mouse, it normalized the recipient's blood sugar for more than a year without requiring any immune suppressing medication at all, because the transplanted tissue, despite having developed inside a rat, was genetically identical to the mouse receiving it.

This experiment, and others like it, represent one of the most scientifically consequential and ethically debated frontiers in modern chimera research: the deliberate creation of chimeric animals in the laboratory, built specifically to advance medicine, and increasingly, to address one of medicine's most persistent and lethal shortages, the lack of transplantable human organs.

The technique underlying this research, called blastocyst complementation, works by exploiting a fundamental principle of early embryonic development. A blastocyst is an early stage embryo, formed just days after fertilization, containing a small cluster of cells with the potential to become any tissue type in the developing organism. If researchers use gene editing tools to disable the genetic instructions responsible for building a specific organ, the resulting embryo develops normally in every other respect but arrives at the developmental stage where that organ should begin forming with no cells present to build it. If donor stem cells from a different source, potentially from an entirely different species, are introduced into that embryo before this critical developmental window closes, those donor cells can step in and fill the vacant developmental niche, building the missing organ using the donor's genetic instructions while the rest of the animal develops using its own original genome. The resulting adult animal is a chimera in the most direct biological sense of the term: one individual, built from two genetically distinct populations of cells, with the vacancy and filling process precisely controlled by researchers rather than occurring through the largely uncontrolled biological accident that produces naturally occurring human chimerism.

The mouse and rat pancreas experiment described above represents a demonstration of principle within a single research approach, but the broader ambitions of this field extend considerably further, toward what researchers in the field describe as one of the most promising potential solutions to a genuinely dire and persistent medical crisis: the severe worldwide shortage of organs available for human transplantation. Thousands of patients die each year while waiting for a kidney, liver, heart, or other organ that never becomes available in time, and the gap between the number of patients who need organs and the number of donated organs that become available shows no sign of narrowing through conventional donation programs alone.

The proposed solution research groups are actively pursuing, and have made substantial documented progress toward, involves using precisely this blastocyst complementation technique to grow human organs inside a different, larger animal host, most commonly pigs, whose organ size and general physiology are considerably closer to human scale than mice or rats. Researchers have used gene editing tools, including the CRISPR gene editing system, to engineer pig embryos lacking the capacity to develop specific organs or tissue types on their own, then introduced human stem cells into these gene edited embryos in the hope that the human cells would fill the developmental gap, building a human organ, grown to full human relevant scale, inside a pig host animal, which could then theoretically be harvested for human transplantation once the animal reached appropriate maturity.

Documented research along these lines has achieved genuine, if still partial, success. Scientists working to produce chimeric pigs carrying human derived skeletal muscle tissue engineered pig embryos with the genetic pathways responsible for skeletal muscle development specifically disabled, then introduced human induced pluripotent stem cells, a category of adult cells reprogrammed in the laboratory to regain the broad developmental potential of embryonic cells, into these gene edited pig embryos. The resulting chimeric pigs, when the procedure succeeded, developed viable, ambulatory bodies with functional skeletal muscle tissue substantially derived from the introduced human cells, an outcome that would have sounded like speculative science fiction to most working biologists only a generation earlier.

This research trajectory, while scientifically significant, has generated substantial and serious ethical debate within both the scientific community and the broader public, debate that researchers actively working in the field have generally acknowledged and engaged with directly rather than dismissing. The central ethical concern running through nearly all serious discussion of human animal chimera research involves the possibility, whether intended or genuinely unintended, that introduced human stem cells might not confine themselves to the specific targeted organ or tissue type researchers intend to grow, but might instead migrate to and contribute meaningfully to other tissues in the host animal, most critically the developing brain or the reproductive organs responsible for producing eggs or sperm.

The concern regarding brain tissue involves the unsettling possibility that a chimeric animal engineered for an entirely different purpose, such as growing a human pancreas or human skeletal muscle, might inadvertently develop a brain containing a meaningful proportion of human derived neurons, raising difficult and largely unprecedented questions about whether such an animal might develop some degree of humanlike cognition or awareness, a possibility that remains highly uncertain given current scientific understanding of how neural tissue origin relates to cognitive function, but one that responsible researchers in the field have taken seriously enough to build specific safeguards against. Published research in this area has documented that in certain chimeric mouse models created using human stem cells, human derived cells have been found to contribute measurably to the animal's brain tissue, in at least one well documented line of research accounting for roughly one percent of all neurons present, alongside contributions to other organs including the liver and thymus. This finding, while representing a fraction of overall brain tissue clearly insufficient to suggest anything approaching human level cognition in the resulting animals, was significant enough that the researchers involved subjected the work to particularly rigorous ethical review specifically because of the theoretical implications of introducing human derived neural tissue into an animal's developing brain, even at a relatively modest proportion of the total.

A parallel and similarly serious ethical concern involves germline contribution, meaning the possibility that introduced human stem cells might migrate into the developing reproductive organs of the host animal and contribute to the production of that animal's eggs or sperm. Should this occur, and should two such chimeric animals subsequently breed, there exists a theoretical possibility of producing offspring carrying human derived genetic material passed down through actual reproduction, a scenario that researchers in the field have generally regarded as ethically unacceptable and have worked specifically to prevent through careful experimental design, typically by using animal hosts and experimental conditions that minimize any realistic possibility of human cell contribution reaching reproductive tissue, and by preventing chimeric research animals from being used for breeding under any circumstances.

Bioethicists writing on this subject have articulated concerns that extend beyond these specific technical worries about brain and germline contribution toward a broader philosophical unease regarding what some describe as the blurring or crossing of species boundaries, and what this blurring might mean for how society understands human dignity and the distinctiveness of human identity relative to other animal species. Critics working from a range of ethical and religious perspectives have argued that the deliberate creation of organisms containing substantial human genetic material embedded within an animal body represents a category of biological manipulation that deserves particular caution, independent of whatever specific practical concerns about cognition or reproduction might apply to any individual research project, because it disrupts a conceptual boundary between human and animal that carries significant moral and cultural weight beyond its purely biological dimensions. Some bioethicists have gone further, arguing that specific newer techniques involving certain categories of human stem cells carry ethical concerns serious enough that this particular line of research should not proceed at all, reflecting a genuine and unresolved disagreement within the bioethics community about where, precisely, the acceptable boundaries of this research program should be drawn.

This unresolved disagreement plays out in concrete terms through the regulatory and funding decisions research institutions and government agencies must actually make, decisions that necessarily require drawing specific lines even in the absence of full ethical consensus about where those lines ideally belong. Research institutional review boards, the committees responsible for approving and overseeing experiments involving human biological material, have developed increasingly detailed and specific protocols governing exactly which categories of human animal chimera research may proceed and under what conditions, typically distinguishing sharply between research using more limited human stem cell types with restricted developmental potential and research using more broadly capable stem cell types carrying a theoretically greater risk of contributing to brain or germline tissue. These review processes generally require researchers proposing human animal chimera experiments to submit detailed plans specifically addressing how they intend to monitor and, if necessary, terminate a developing chimeric embryo or animal should evidence emerge of unexpectedly extensive human cell contribution to sensitive tissue types, a regulatory safeguard that reflects the broader scientific community's attempt to proceed with this ethically significant research in a manner that takes seriously, rather than dismisses, the legitimate concerns raised by its critics.

Proponents of continued research, including many of the scientists actively conducting this work, generally respond to these concerns not by dismissing them, but by emphasizing the carefully controlled experimental conditions used specifically to minimize the risk of unintended brain or germline contribution, and by pointing toward the substantial and demonstrated potential benefit of this research: a genuine, scientifically grounded pathway toward addressing a organ shortage crisis that currently costs thousands of human lives annually, a benefit proponents argue deserves serious weight in any ethical calculation alongside the legitimate concerns critics have raised. Government research funding agencies in several countries, including the United States National Institutes of Health, have supported this research area with public funding following extensive internal ethical review processes, while also maintaining specific funding restrictions on the more ethically fraught subcategories of this research, particularly regarding experiments with the greatest theoretical potential for unintended brain or germline contribution.

The scale of the organ shortage crisis that motivates this research deserves to be stated plainly, since the abstract ethical debate outlined above can otherwise obscure just how significant the human cost of organ scarcity actually is. National transplant registries in multiple countries have documented waiting lists numbering in the tens of thousands of patients at any given time, with a substantial proportion of those waiting for a kidney specifically, the organ for which demand most dramatically outpaces the available supply of donated organs from both deceased and living donors combined. Patients on these waiting lists die at a measurable and significant rate every year, not because effective treatments do not exist for their underlying conditions, but specifically because no compatible organ becomes available in time. This is the specific, concrete human stake that researchers pursuing human animal chimera organ growth research point to when arguing that the ethical concerns raised by critics, while genuinely serious and deserving of careful ongoing consideration, must be weighed against an equally serious and equally concrete cost associated with not pursuing this research, or with pursuing it more slowly and more cautiously than current organ shortage statistics might otherwise justify.

Alongside blastocyst complementation research aimed at growing entire organs, a related and more immediately clinically advanced field called xenotransplantation has already reached the stage of active, if still highly experimental, use in human patients, offering a somewhat different technical approach to the same underlying organ shortage problem. Rather than growing an entire organ from human cells inside an animal host through blastocyst complementation, xenotransplantation research has focused on genetically modifying pig organs directly, using gene editing techniques to remove or alter specific pig genes responsible for triggering severe immune rejection in human recipients, with the goal of producing a pig organ, remaining entirely pig derived rather than substantially human cell derived, that a human immune system can nonetheless tolerate reasonably well after transplantation. This approach has already progressed to the point of gene edited pig kidneys being transplanted into human patients, including, in a small number of highly documented and closely monitored cases, patients who had been declared clinically deceased, a research approach used specifically to study organ function and immune response in a real human physiological environment before proceeding toward transplantation in living patients.

Humanized mouse models represent a third, considerably longer established and less ethically contested category of deliberately created chimera, widely used throughout biomedical research for decades. These models are typically created by introducing human cells, often human immune cells or human tumor cells, into mice that have been genetically modified or otherwise treated to lack a functional immune system of their own, preventing the mouse's immune system from rejecting the introduced human cells. The resulting chimeric mice, carrying a functioning human immune system or a growing human tumor within an otherwise normal mouse body, provide researchers with an invaluable and widely used experimental platform for studying human specific diseases and testing potential human treatments in a living organism, without the ethical and practical barriers involved in conducting equivalent experiments directly in human patients at an early, unproven stage of research.

The scientific value of humanized mouse models extends across an unusually broad range of medical research applications, reflecting just how central this particular chimeric technique has become to modern biomedical science despite receiving far less public attention than the more visually striking and ethically fraught human pig organ research discussed earlier in this chapter. Researchers studying infectious diseases that affect only human cells, and that therefore cannot be studied directly in ordinary, non humanized laboratory animals, have relied on humanized mouse models to observe how such pathogens actually behave within a living immune system, insight that would otherwise require testing directly in human volunteers at a stage of research when the safety and efficacy of a candidate treatment remains genuinely unknown. Cancer researchers have similarly relied heavily on a specific category of humanized mouse model in which a sample of a patient's own tumor is transplanted directly into an immune deficient mouse, allowing the tumor to continue growing within a living organism where researchers can test how it responds to different candidate treatments before those treatments are ever administered to the actual human patient whose tumor was used to create the model, an approach that has become an increasingly important tool within the broader movement toward personalized cancer treatment tailored to the specific biological characteristics of an individual patient's disease.

Immunologists studying how the human immune system develops and functions have found humanized mouse models particularly valuable for a distinct reason connected directly back to the immune tolerance concepts discussed earlier in this chapter and in the previous chapter's discussion of microchimerism. Because these models involve introducing human immune cells into a mouse at a stage where the mouse's own immune system has been suppressed or genetically prevented from developing normally, researchers can observe how human immune cells behave and mature within a living organism largely free from competing immune signals, offering insight into human immune development that would be considerably more difficult to obtain through observation of human patients alone, given the practical and ethical limitations involved in directly manipulating and observing human immune development in living human subjects.

Long before any laboratory deliberately engineered a chimera for medical research, nature had already been producing them spontaneously and repeatedly in a far more visible and far more commonly encountered setting than any research facility: the household cat. Chimerism in cats offers what is very likely the most accessible, most photographed, and most widely recognized visual demonstration of the phenomenon available anywhere, precisely because feline chimerism frequently produces a strikingly obvious physical signature that requires no laboratory equipment whatsoever to notice, only a pair of eyes.

A true chimera cat typically displays what owners and veterinary geneticists alike describe as a split face or two toned coat, a sharp, often remarkably straight dividing line running down the center of the animal's face or body, with distinctly different fur colors on either side, sometimes accompanied by heterochromia, in which each eye carries a different color, one commonly appearing blue while the other appears green or amber. The mechanism producing this pattern mirrors precisely the tetragametic fusion process described earlier in this book in the context of human chimerism: two separately fertilized feline embryos, which would otherwise have developed into two separate kittens, merge at an early developmental stage into one continuing embryo, carrying forward two genetically distinct cell lineages that, as the resulting single kitten develops, distribute themselves unevenly across its body, producing the visually dramatic split coloration that has made certain individual chimera cats into minor internet celebrities, their photographs circulated widely specifically because the phenomenon looks so implausible that many viewers initially assume the images have been digitally manipulated.

It is worth pausing specifically to correct a persistent and extremely common point of confusion surrounding feline chimerism, because the confusion itself illustrates something valuable about how genetics can produce superficially similar outward appearances through entirely different underlying mechanisms. The vast majority of two toned or patchily multicolored cats that people encounter, including the very familiar tortoiseshell and calico coat patterns found throughout ordinary domestic cat populations worldwide, are not chimeras at all, despite bearing a superficial resemblance to true chimeric coloration. These far more common patterns arise instead through a process called X inactivation, which operates on an entirely different genetic principle. Female cats carry two X chromosomes in every cell, and different color related genes can be carried on each of these two X chromosomes. Early in a female embryo's development, one of the two X chromosomes in each individual cell is randomly and permanently deactivated, a process that happens independently in each cell rather than uniformly across the whole body, meaning that some patches of the resulting cat's skin will express the color gene carried on one X chromosome, while other patches express the color gene carried on the other X chromosome. The result, a mottled, multicolored coat, can look remarkably similar to true chimerism at a casual glance, but it involves only one single genetic identity throughout the entire animal, split in its outward gene expression by X inactivation rather than by the fusion of two genuinely separate embryos, as occurs in true chimerism.

Distinguishing a genuine chimera cat from an ordinary tortoiseshell or calico cat displaying X inactivation patterning generally requires actual genetic testing, since the visual patterns the two mechanisms produce can overlap substantially. Veterinary geneticists who study the distinction have noted specific circumstances that raise a stronger likelihood of true chimerism, most notably in male cats displaying tortoiseshell or calico coloration, since this particular coat pattern normally requires two different X chromosomes, and male cats ordinarily carry only one X chromosome alongside one Y chromosome. A male cat displaying full tortoiseshell or calico patterning is therefore genetically unusual by definition, and while some such cats carry an extra X chromosome, a condition analogous to Klinefelter syndrome in humans, researchers studying the phenomenon have suggested that a meaningful proportion of male calico and tortoiseshell cats may in fact be true chimeras, carrying two distinct cell lineages that happen to include both a male determining and a color determining genetic contribution, distributed across the body in a pattern that produces the classic tortoiseshell or calico look despite the cat's overall male sex.

Documented cases of confirmed chimerism in cats extend beyond simple coat coloration into more genuinely complex genetic territory as well, including cases in which researchers have identified two entirely different coat colors, one a standard, undiluted color and the other a diluted, muted variant, present simultaneously in a single cat's fur and reproductive tissue in a pattern that could only be explained by two separately fertilized eggs having fused into one developing kitten, since no single genetic lineage operating alone could produce that particular combination of traits. Feline genetics researchers examining such cases have used exactly the same investigative logic employed in human chimerism cases like Karen Keegan's, tracing genetically inconsistent traits across multiple body regions, including testicular tissue in male cats used for breeding, to confirm that two distinct cell lineages were present and unevenly distributed throughout the animal's body.

Veterinary researchers have also noted that chimeric cats and dogs, much like their human counterparts, can experience genuine health complications stemming directly from their dual genetic makeup, distinct from the largely cosmetic coat and eye color effects that make chimerism visually recognizable in the first place. Because the immune system develops its foundational sense of which cells count as native, native self versus foreign material during an animal's earliest development, a chimeric animal's immune system may, in some documented cases, only fully accept one of its two cell lineages as legitimately its own, treating tissue derived from the second lineage with a degree of ongoing immunological suspicion that can manifest as autoimmune type complications, a parallel to the immune irregularities described by human chimera Taylor Muhl in an earlier chapter of this book.

Feline and canine chimerism, precisely because it is visually striking, widely documented by pet owners and breeders long before genetic testing was affordable or widely available, and now increasingly confirmed through accessible commercial pet DNA testing services, offers something that human chimerism research has historically lacked: a reasonably large, ongoing, and continuously growing body of casually documented cases, driven not by dedicated research funding but by ordinary pet owners noticing something visually unusual about their own animals and pursuing an explanation, sometimes purely out of curiosity, sometimes because a breeder needs to understand an unexpected coat pattern appearing in a litter. This accumulated body of case documentation, while never intended as formal scientific research, has nonetheless meaningfully advanced general public and veterinary understanding of how chimerism actually presents in a living, breathing animal, offering an accessible and visually intuitive entry point into a phenomenon that, in its human form, remains considerably harder for most people to picture or fully grasp.

What unites all of these deliberately engineered chimera research programs, from decades old humanized mouse models to the newest and most ethically contested human pig organ research, alongside the entirely spontaneous, naturally occurring chimerism documented extensively in domestic cats and dogs, is a shared scientific insight directly connected to the naturally occurring human chimerism examined throughout the earlier chapters of this book: that a single body, built from genetically distinct populations of cells working alongside each other, is not merely a biological curiosity or a rare accident of twin embryo fusion, but a genuinely functional biological arrangement, one that nature has produced spontaneously across multiple species, from the household cat curled up on a windowsill to Lydia Fairchild and Karen Keegan, and one that human researchers have learned, with steadily increasing precision and steadily expanding ambition, to deliberately construct for their own scientific and medical purposes as well.

The next chapter turns from laboratory science back toward mythology and cultural history, examining the ancient origins of the word chimera itself, and the surprisingly long tradition, stretching across many cultures and many centuries, of human storytelling grappling with the same unsettling idea that modern genetics has now confirmed as biological fact: that a single body might, under the right and exceedingly rare circumstances, contain more than one distinct identity at once.

CHAPTER 10
The Ancient Chimera

In the mountainous region the ancient Greeks called Lycia, in what is today southwestern Turkey, farmers told stories of a creature that scorched their fields and slaughtered their livestock, a monster no ordinary warrior dared approach. Homer, writing in the Iliad sometime around the eighth century before the common era, described this creature as being of divine origin rather than born of ordinary animal parents, a being with the front half of a lion, the tail of a serpent, and, rising from its middle, the head and body of a goat, through which it breathed a terrible and unrelenting fire. Hesiod, another of the earliest Greek poets whose work survives, described the beast similarly in his own account, agreeing on the essential combination of lion, goat, and serpent, though later writers and artists would render the exact arrangement of these three elements with some variation, sometimes placing the goat's head sprouting directly from the creature's back rather than fully integrated into its central body.

The Greeks named this creature Chimera, and its story, as preserved across multiple ancient sources, involves the hero Bellerophon, a young man of royal blood who had been exiled following a serious, though variously described, crime, and who found himself, through a chain of court intrigue and false accusation involving a scorned queen, sent by the king of Lycia on what was intended to be a suicide mission: to track down and kill the Chimera that had been terrorizing the kingdom's countryside. What the king did not anticipate was that Bellerophon possessed a singular advantage unavailable to any ordinary warrior who might have attempted the same task before him. He had tamed Pegasus, the winged horse born from the severed neck of the slain Gorgon Medusa, and riding Pegasus into the sky above the Chimera's reach, Bellerophon was able to attack the fire breathing monster from a position it could neither reach nor effectively defend against. According to the most commonly repeated version of the myth, Bellerophon finished the creature by attaching a lump of lead to the tip of his spear and thrusting it directly into the Chimera's fire breathing mouth, where the intense heat melted the lead, which then flowed down the creature's throat and killed it from within.

This story, more than two and a half thousand years old, might seem to have little to say to a book about modern genetics, and in most respects it does not. The Chimera of Greek myth was not, in any way its original storytellers intended, an attempt to describe a biological phenomenon. It was, most historians of mythology agree, likely rooted in something considerably more mundane: ancient observers in the region around Mount Cragus on the Lycian coast are recorded by later geographers as having associated the myth with a real geological feature, a mountain from which flames were said to emerge, likely from natural gas seeping through rock and igniting on contact with air, a phenomenon that persists in that same region into the present day and that ancient peoples, lacking any concept of underground natural gas deposits, quite reasonably explained through the more vivid and narratively satisfying framework of a monstrous, fire breathing beast inhabiting the mountainside.

And yet, despite this almost certainly mundane geological origin, the specific imaginative shape the ancient Greeks gave their monster, a single creature built simultaneously from the complete forms of three entirely different animals, coexisting rather than blending into some smoothed over hybrid form, turned out to be a remarkably precise conceptual match for the biological phenomenon this book has spent its previous chapters describing. This is worth pausing on, because it is easy to assume that scientists simply grabbed a convenient, colorful mythological name off the shelf when they needed a word for chimerism, in the same casual way that countless scientific terms borrow loosely and somewhat arbitrarily from classical mythology. But the fit between the ancient Chimera and the modern biological chimera is closer and more conceptually apt than most such borrowings.

Consider what the Greek Chimera actually was, according to its earliest and most consistent textual descriptions. It was not a creature with, say, features blended evenly across its entire body, some vague composite smear of lion, goat, and snake characteristics distributed uniformly throughout a single unified anatomy. It was, in Homer's own words, lion in its front portion, serpent in its hind portion, and goat specifically in its middle, breathing fire through the goat's own separate head. Each section of the creature's body remained recognizably and completely itself, a genuine lion's portion, a genuine serpent's portion, a genuine goat's portion, coexisting within one physical form rather than merging into some new, intermediate creature that was part lion, part goat, and part serpent in every part of its body simultaneously. This distinction, between complete coexistence of distinct forms and blended hybridization, maps with real precision onto the biological difference between chimerism and ordinary genetic inheritance.

An ordinary child, inheriting genetic material from both a mother and a father, is a blend, a genuine hybrid in the most literal genetic sense, carrying a single unified genome built from contributions of both parents combined together into one new, intermediate genetic identity present uniformly in every cell of that child's body. This is not what happens in chimerism. A chimeric individual like Lydia Fairchild or Karen Keegan does not carry some blended, averaged genetic identity distributed evenly throughout her body. She carries two entirely separate, entirely complete, entirely distinct genetic identities, each one fully itself, coexisting in different proportions across different regions and tissues of a single physical body, precisely as the mythological Chimera carried a complete lion's anatomy, a complete goat's anatomy, and a complete serpent's anatomy, each remaining entirely itself, within one physical creature.

This is, when the parallel is laid out explicitly, a genuinely elegant conceptual match, arguably closer and more apt than the vast majority of scientific terms borrowed from classical mythology, many of which capture only the loosest, most superficial thematic resemblance to whatever phenomenon they eventually came to describe. The scientists who first applied the term chimera to human tetragametic chimerism, credited in the historical record to research beginning in the middle of the twentieth century, appear to have recognized this deeper structural resonance rather than simply reaching for an evocative sounding classical reference.

The mythological Chimera also carries a notable detail regarding its origin that resonates, whether by genuine ancient insight or simple coincidence, with the biological mechanism underlying modern chimerism. According to the most widely repeated ancient genealogy, recorded by both Hesiod and later mythographers, the Chimera was the offspring of Typhon, a monstrous storm giant described as possessing a hundred serpent heads and voices capable of producing every sound in the animal kingdom, and Echidna, herself described as half woman and half serpent, a mother of numerous other legendary monsters throughout Greek mythology including Cerberus, the three headed hound that guarded the entrance to the underworld, and the Sphinx, the riddling half human, half lion creature that famously tormented the city of Thebes. This genealogy positions the Chimera, from its very conception in the mythological tradition, as inherently and originally composite, born already carrying the blended monstrous heritage of two entirely non uniform parents, rather than beginning as some simple, singular creature that later somehow acquired its composite nature through injury, magic, or transformation. The Chimera, mythologically speaking, was never meant to be one thing. Multiplicity was built into its origin from the very beginning, precisely as tetragametic chimerism in humans traces back to a moment of origin, the fusion of two separately fertilized embryos, that occurs before any single, uniform biological identity had the chance to fully establish itself.

The endurance of the Chimera as a cultural touchstone extends well beyond its original mythological context and its subsequent adoption as a scientific term. Roman writers, including Virgil in the Aeneid, continued to reference the creature centuries after Homer and Hesiod first recorded its story, placing it among the assembled monsters guarding the entrance to the underworld. The image of the Chimera appears repeatedly across ancient Greek and Roman visual art, most notably on pottery and in sculpture, including one particularly famous bronze sculpture discovered in the Italian city of Arezzo, dated to the fifth century before the common era, depicting the creature with wounds already inflicted, presumably representing the moment of its defeat by Bellerophon, a work that remains among the most celebrated surviving examples of ancient Etruscan bronze work. The term chimera itself gradually expanded well beyond its original mythological referent to become, in ordinary English usage that persists into the present day, a general word for any fanciful, wildly improbable, or ultimately illusory idea or hope, as in a phrase like "a chimerical scheme," reflecting the creature's essential unreality and its combination of elements that, outside of myth, simply do not belong together.

This broader linguistic legacy makes the specific scientific adoption of chimera as a precise biological term somewhat unusual among mythologically derived scientific vocabulary, because it required scientists to draw specifically and deliberately on the creature's original, literal anatomical description, its composite structure built from genuinely distinct and complete parts, rather than simply borrowing the word's later, looser cultural connotation of fancifulness or impossibility. In this sense, the scientific term chimera represents something of a return to the myth's original, most literal meaning, stripped of the centuries of accumulated figurative usage that had grown up around it in ordinary language, reapplied with real precision to a genuine biological phenomenon that ancient storytellers, describing a mountain that breathed fire, could never have anticipated their monster would eventually come to name.

Beyond the specifically Greek tradition, numerous other cultures across history and geography developed their own mythological traditions involving composite creatures built from the combined forms of multiple distinct animals, suggesting that the underlying conceptual fascination, and perhaps unease, with a single body containing multiple complete identities reflects something broader in human imagination than one specific cultural tradition. Ancient Egyptian religious iconography prominently featured deities combining human and animal forms, including Anubis with the head of a jackal and Sobek with the head of a crocodile, though these figures generally represent a somewhat different conceptual category than the Greek Chimera, typically understood as deliberate divine forms rather than monstrous accidents of unnatural combination. Ancient Mesopotamian art and mythology similarly featured composite creatures, including winged bulls with human heads positioned as protective guardian figures at palace entrances. The sphinx, appearing in both Egyptian and Greek tradition though with notably different characteristics and narrative roles in each, likewise represents a composite creature combining human and animal elements that predates and parallels the specific Chimera myth.

What these varied and geographically dispersed traditions collectively suggest is that the imaginative territory chimerism now occupies as a confirmed biological reality, a single body built from more than one complete identity, has occupied human storytelling and religious imagination across a remarkably wide range of independent cultural traditions for thousands of years before any scientist could confirm that such an arrangement was biologically possible. Modern genetics did not introduce humanity to the unsettling and fascinating idea of a body containing more than one self. It confirmed, in the most literal and clinical terms imaginable, that an idea humanity's storytellers had already been exploring for millennia was not merely a productive metaphor or a useful narrative device, but, in an exceedingly rare and specific set of circumstances, biological fact.

The persistence of composite creature mythology extends into considerably more recent cultural history as well, offering a useful reminder that this particular imaginative fascination did not simply fade away once ancient religious traditions gave way to more modern secular storytelling. Medieval European bestiaries, illustrated manuscripts cataloguing real and imagined animals for moral and educational purposes, continued to feature composite creatures drawing directly on classical mythological sources, including the Chimera itself, which appears in various medieval artistic and literary contexts long after the decline of the classical civilizations that first told its story. Heraldry, the elaborate visual system of symbols and designs used across medieval and early modern Europe to identify noble families and their lineages, similarly drew heavily on composite mythological creatures, including griffins combining lion and eagle characteristics and wyverns combining serpent and dragon characteristics, reflecting a continued cultural fascination with creatures built from more than one complete animal identity that persisted across an enormous span of European cultural history entirely independent of any biological knowledge that might eventually make such creatures scientifically plausible in a different, cellular sense.

This same fascination continues into contemporary popular culture in forms considerably more removed from ancient mythology than medieval heraldry, yet clearly drawing on the same underlying narrative appeal. Modern fiction, film, and other media continue to feature composite or hybrid creatures and characters combining multiple distinct identities within a single form, a narrative pattern that storytelling across cultures and centuries has returned to again and again, suggesting that whatever makes the idea of a single body containing more than one complete identity narratively compelling reflects something durable and widely shared in human imagination, rather than a passing fascination specific to any one culture or historical period. Modern genetics, in confirming that this pattern occasionally occurs as literal biological fact, has not so much introduced a new idea into this long storytelling tradition as revealed that one small, rare corner of it was true all along.

The chapter that follows turns from this ancient mythological and cultural context toward a more directly modern and personal set of questions: what chimerism actually means, philosophically and psychologically, for the individuals who carry it, and what it suggests more broadly about the nature of personal identity, individuality, and selfhood for everyone, chimeric or not.

CHAPTER 11

Two Minds, One Body

Picture two fraternal twin embryos, each carrying its own complete and distinct genetic identity, floating within the same early uterine environment during the first days after fertilization. Now imagine that instead of continuing to develop as two separate embryos, eventually born as two separate people, they merge, days later, into a single developing organism that continues on to become one adult human being. That resulting person will grow up, form relationships, develop opinions, fall in love, pursue a career, and experience the entire continuous, unified sense of being one single self that every person carries through life, entirely unaware, in the vast majority of cases, that their body actually originated from what would have been two people rather than one.

Who, in that scenario, actually exists? Is the resulting adult one person who happens to carry two genetic lineages, or is that adult, in some more troubling sense, the fusion of two people who each, individually, ceased to exist as separate potential selves the moment their embryos merged? Did two potential lives become one, with something genuinely lost in the merger, or was there only ever going to be one life all along, one that simply happened to be built from materials that briefly existed as two?

These questions sound, at first pass, like the kind of abstract thought experiments philosophers construct for classroom discussion, disconnected from any real world stakes. Chimerism reveals that they are not abstract at all. They describe, in perfectly literal terms, what actually happened to Lydia Fairchild, to Karen Keegan, and to Taylor Muhl, each of whom lives, right now, as a single continuous person with a single continuous sense of self, despite having originated from the physical fusion of what would otherwise have been two genetically distinct fraternal twin siblings.

Philosophers working in the field of personal identity, the branch of philosophy specifically concerned with what makes a person the same individual over time and what distinguishes one person from another, have generally approached questions of identity through frameworks developed with a background assumption that each person's identity traces back cleanly to a single, unified biological origin: one sperm, one egg, one zygote, one continuous developmental line leading to one adult. Chimerism does not fit neatly into this framework, because it involves precisely the kind of biological origin story, two separate zygotes merging into one developmental line, that most traditional philosophical accounts of personal identity never anticipated needing to address.

Some philosophers who have engaged specifically with the implications of chimerism argue that the resulting individual should be understood as a genuinely single person from the moment of fusion onward, on the grounds that personal identity, properly understood, tracks continuous psychological and biological development over time rather than tracking genetic uniformity specifically. On this view, what makes Lydia Fairchild one continuous person is not that every cell in her body shares an identical genetic origin, but that she has experienced one continuous stream of consciousness, one continuous body developing and aging together as an integrated whole, and one continuous set of memories and relationships extending back as far as any human memory extends. The specific genetic composition of her cells, on this account, is simply not the relevant criterion for determining personal identity, any more than the fact that essentially all of the atoms making up any adult human body will have been entirely replaced multiple times over the course of that person's life, through ordinary cellular turnover and metabolism, threatens the intuitive judgment that a person remains the same continuous individual across those decades of complete physical replacement.

Other philosophers, approaching the same question from a different angle, have focused less on defending a confident answer to what chimerism means for personal identity and more on using chimerism as a genuinely useful test case for exposing hidden assumptions embedded within existing philosophical frameworks that most previous theorists never had reason to examine closely, because the biological phenomenon that would challenge those assumptions was not widely known or discussed until relatively recently. Chimerism, on this view, functions less as a phenomenon requiring an immediate philosophical resolution and more as what philosophers sometimes call an intuition pump, a scenario specifically useful for revealing where our ordinary, everyday concept of personal identity, built up through experience with the overwhelming majority of ordinary, non chimeric people, quietly relies on assumptions about genetic uniformity that turn out not to be strictly necessary, or even always strictly true, once genuinely unusual cases like tetragametic chimerism are taken seriously.

The question of whether something is genuinely lost when two potential twin embryos fuse into one deserves particular care, because it touches on emotionally significant territory connected to how people think and feel about pregnancy loss more broadly, a subject already touched on in the earlier discussion of vanishing twin syndrome. It would be a mistake, and arguably a philosophically confused one, to describe chimeric fusion in language suggesting that one twin somehow died so that the other could live, because this framing implies a level of developed, separate personhood in each embryo at the moment of fusion that the actual biological facts do not support. The fusion occurs within the first two weeks after fertilization, at a developmental stage before either embryo has formed anything resembling a nervous system, before any capacity for consciousness, sensation, or experience of any kind could plausibly exist. Whatever moral or metaphysical status one assigns to a very early embryo, a question connected to some of the most contested debates in bioethics and one this book does not attempt to resolve, the specific timing of chimeric fusion places it well before the developmental milestones that most ethical and philosophical frameworks, across a wide range of otherwise differing views, treat as relevant to questions of individual personhood or the capacity for experience.

What can be said with somewhat more confidence, because it draws on the direct testimony of people actually living with confirmed chimerism rather than abstract philosophical speculation, is how chimeric individuals themselves have described the psychological experience of learning about their own condition. Taylor Muhl's account, discussed in an earlier chapter, emphasizes relief rather than existential distress, a sense of finally receiving a coherent explanation for physical and immunological experiences that had previously seemed disconnected and inexplicable. This response pattern, relief and clarity rather than a destabilizing crisis of identity, appears consistently across the limited but genuine testimony available from confirmed chimeric individuals who have spoken publicly about their diagnosis, and it offers a meaningful, if necessarily limited, real world data point suggesting that the philosophical unease chimerism can provoke in outside observers contemplating the phenomenon in the abstract does not necessarily translate into equivalent psychological distress for the actual individuals living with the condition day to day.

This makes a certain intuitive sense once considered carefully. A chimeric individual's lived, subjective experience of selfhood does not actually change in any way once they learn their diagnosis. Lydia Fairchild's sense of being herself, of having lived one continuous life, of being unambiguously the mother who carried and raised her own children, was never in question from her own first person perspective at any point during her legal ordeal. What was in question was how outside institutions, relying on an incomplete and ultimately misleading genetic test, chose to interpret evidence about her body. The diagnosis of chimerism did not introduce a new form of internal identity confusion into Lydia's life. It resolved an external, institutional confusion that had been imposed upon her from outside, restoring a straightforward alignment between the biological facts and the lived reality she had never doubted.

It is worth contrasting this pattern with a genuinely different category of identity related medical and psychological experience, specifically to sharpen exactly what chimerism does and does not resemble. Dissociative identity disorder, a recognized psychiatric condition involving the presence of two or more distinct personality states within a single individual, is sometimes invoked, whether by curious readers or by less careful popular commentary, as a loose psychological analogy to chimerism, given that both conditions involve some form of multiplicity coexisting within a single body. This comparison, however, rests on a fundamental confusion between two entirely unrelated phenomena operating at completely different levels of biological and psychological organization. Chimerism is a condition of cellular genetics, describing the coexistence of two distinct genetic lineages within different tissues of a single physical body, with no direct bearing whatsoever on that individual's psychological experience of selfhood, which, as the testimony and evidence discussed throughout this chapter indicates, remains entirely singular and continuous regardless of the person's chimeric status. Dissociative identity disorder, by contrast, is a psychiatric condition involving genuine disruptions to psychological continuity and self experience, typically understood by clinicians who treat the condition as connected to early and severe childhood trauma, and it has no established genetic or cellular chimerism component whatsoever. Conflating the two, beyond simply being scientifically inaccurate, risks trivializing the genuine psychological distress experienced by people living with dissociative identity disorder, while simultaneously overstating and misrepresenting the actual, generally psychologically unremarkable lived experience of confirmed chimeric individuals like Taylor Muhl. This book has aimed throughout to keep this distinction clear, treating chimerism strictly as the biological, cellular phenomenon it actually is, rather than allowing it to be casually or inaccurately merged with entirely separate psychological or psychiatric territory that happens to share a superficial vocabulary of multiplicity.

This distinction, between the philosophical puzzle chimerism poses when contemplated abstractly and the comparatively unremarkable lived experience it produces for actual chimeric individuals, connects to a broader and more general point about personal identity that extends well beyond chimerism specifically. The felt, subjective experience of being a single, continuous self does not, for any person, chimeric or otherwise, actually depend on continuously verifying or even being aware of the underlying biological facts that philosophers debate when constructing theoretical accounts of personal identity. Nobody walks through ordinary daily life running mental checks to confirm that their cells share a uniform genetic origin before feeling confident in their own selfhood. The sense of being oneself is something closer to a continuously and automatically generated psychological experience, grounded in memory, embodiment, and ongoing relationship with the world, rather than something that would collapse or fragment upon learning an unexpected fact about one's own cellular genetic composition.

Twin studies, an entirely separate and much more extensively developed area of psychological and behavioral genetics research, offer a useful point of comparison and contrast here, because they approach questions of identity and individuality from research involving actual living twin pairs rather than the fused, singular individuals chimerism produces. Decades of twin research, comparing identical twins who share essentially all of their genetic material with fraternal twins who share only the ordinary degree of genetic similarity found between any siblings, have consistently found that even genetically identical twins, raised in the same household under highly similar environmental conditions, develop distinct personalities, distinct preferences, and a robust, unambiguous sense of being separate individuals from one another, despite their near total genetic identity. This research finding, that genetic identity alone does not determine or even strongly predict a merged or confused sense of personal identity between two genetically identical people, offers an interesting point of contrast with chimerism, where the situation runs in the opposite direction: two genetically distinct lineages coexisting within what nonetheless develops into one entirely unified, singular sense of self. Together, these two bodies of evidence, identical twins experiencing themselves as fully separate individuals despite genetic identity, and chimeric individuals experiencing themselves as fully unified individuals despite genetic duality, suggest fairly strongly that personal identity, at least as it is actually lived and experienced rather than as it is sometimes theorized about in the abstract, tracks something considerably more tied to continuous embodied experience and psychological continuity than it tracks strict genetic uniformity or genetic distinctness in either direction.

This same body of twin research offers a further, related insight worth drawing out explicitly, concerning the specific role the brain plays in grounding personal identity as compared with the rest of the body. Chimeric individuals whose two cell lineages happen to be distributed such that one lineage dominates the brain and nervous system, even if the other lineage is more prominently represented elsewhere in the body, such as in blood, skin, or reproductive tissue, would be expected, based on everything currently understood about the neurological basis of consciousness and psychological continuity, to experience an entirely undivided, singular sense of self tracking that dominant neural lineage, regardless of how the remainder of their body's tissue happens to be genetically composed. This expectation aligns closely with the actual reported experience of confirmed chimeric individuals like Taylor Muhl, whose sense of being one continuous person appears entirely unaffected by the fact that a meaningful proportion of her body's cells, quite possibly extending to portions of her brain itself, though this has not been the specific subject of published testing in her case, derive from a genetically distinct lineage. The evidence available, while admittedly limited in scope given how few confirmed chimerism cases exist to study directly, is consistent with a view in which the specific genetic composition of the brain, whatever its precise details, generates one coherent and continuous stream of experience, and that this coherence does not require, and does not appear meaningfully disrupted by, the coexistence of a second, genetically distinct cell lineage elsewhere in the same body.

None of this fully resolves the deeper philosophical puzzle chimerism raises, and responsible philosophical engagement with the topic should resist the temptation to claim otherwise. The question of what, precisely, grounds personal identity, whether it is fundamentally a biological fact, a psychological fact, a matter of continuous first person experience, or some combination of these and other factors entirely, remains genuinely unsettled territory within academic philosophy, debated actively by specialists who bring far more sustained rigor to the question than any single chapter of a popular nonfiction book focused primarily on the biology and human stories of chimerism could hope to offer. What chimerism does contribute to this long running philosophical conversation is a genuinely rare and clarifying real world case, a scenario philosophers might otherwise have needed to construct entirely hypothetically, that forces careful thinking about assumptions most people, philosophers included, rarely have occasion to examine, because the overwhelming majority of human beings never encounter a situation in which those assumptions are tested against an actual, confirmed instance of biological reality.

A related and equally difficult question concerns how chimerism ought to be discussed with children who are found to carry the condition, or with the family members of someone diagnosed later in life. Genetic counselors working with families affected by a confirmed chimerism diagnosis have generally emphasized the importance of clear, accurate, and reassuring communication, precisely mirroring the emphasis this book has placed throughout on correcting the common but mistaken impression that chimerism somehow compromises or divides a person's identity. A parent who learns that their child is a chimera, whether through an incidental discovery during unrelated medical testing or through investigation of a visible physical marker, is not learning that their child is somehow two children, or that some fundamental aspect of their child's selfhood is in question. They are learning a genuinely interesting and rare fact about their child's cellular biology, one that, based on the accumulated case histories and personal testimony examined throughout this book, need carry no more psychological weight than any other unusual but benign medical finding, provided it is explained accurately and without unnecessary alarm.

This communication challenge extends into the broader medical and media landscape as well. Sensationalized reporting on chimerism cases, of the kind that has occasionally surrounded both Lydia Fairchild's and Karen Keegan's stories in less careful corners of the popular press, tends to lean heavily on language suggesting these women were somehow revealed to be two people, or that their identities were thrown into genuine doubt by their diagnoses, framings that this book has worked throughout its earlier chapters to correct and complicate. Responsible science communication on chimerism, of the kind genetic counselors and researchers who work directly with affected families generally strive to model, treats the condition as a fascinating instance of biological variation deserving of careful, accurate explanation, rather than as an occasion for manufactured existential drama that serves headline writers considerably better than it serves the people actually living with the condition.

The next chapter turns from these philosophical questions back toward a more empirical and practical one: given how rarely chimerism is actually diagnosed, and given how many subtle and easily overlooked forms it can take, how many people might be living, right now, as undiagnosed chimeras without any awareness of their own condition, and what, if anything, might change this going forward.

CHAPTER 12

The Undiagnosed Millions

A thirty four year old man in Washington State took a standard paternity test, expecting the routine confirmation that he was the biological father of his son. The test excluded him. Not partially, not ambiguously, but with the same flat, unequivocal certainty that had initially condemned Lydia Fairchild more than a decade earlier. Confused and troubled by a result that contradicted everything he and his family understood about their own history, the man and his partner tried a different kind of test, a consumer ancestry and genealogy service rather than a dedicated paternity laboratory. That test returned something almost stranger than an outright exclusion. Rather than showing no relationship at all, it suggested the man was not his son's father, but his uncle.

Standard paternity tests, of the kind used in family court and immigration proceedings, typically examine somewhere around fifteen specific genetic markers, a number that is more than sufficient to establish or exclude parentage with overwhelming statistical confidence in the vast majority of ordinary cases. Consumer genetic ancestry services, built for the very different purpose of mapping extended family trees and estimating ethnic ancestry, examine several hundred thousand genetic markers across the entire genome, providing a dramatically more detailed and sensitive picture of any individual's genetic relationships. It was this far more granular consumer test, not the specialized paternity laboratory equipment, that first revealed the specific and highly unusual pattern consistent with chimerism, an outcome the geneticists eventually involved in the case would later describe as "blown away" by, in the words of one Stanford University geneticist consulted on the case, despite this being an area of research these specialists had studied for years.

Investigators, including researchers from a major DNA diagnostics laboratory and geneticists from Stanford University and Massachusetts General Hospital, ultimately determined that the man had originated as one half of a dizygotic, fraternal twin pregnancy. His twin had not survived to birth, but had not simply vanished without a trace either. Cells from the unborn twin had been incorporated directly into the surviving twin's own developing reproductive tissue, meaning that a portion of his sperm carried the genetic signature of the twin who had never been born, rather than his own. When he had a son using sperm carrying his own original genetic line, that child tested, correctly, as his biological son under standard testing. When he had a second son using sperm derived from the incorporated twin's genetic lineage, that child's DNA matched a different genetic profile entirely, one technically belonging to an individual who had never lived outside the womb, producing a result that looked, on a sufficiently detailed genetic test, like an uncle rather than a father.

Physicians involved in investigating the case noted a detail that, in hindsight, offered a visible clue that had gone unrecognized for over three decades of the man's own life: distinctly two toned, patchy skin, pronounced enough that as a child he had reportedly told other children he was a burn victim rather than explain a physical difference nobody around him, including his own family and doctors, had ever correctly identified as a sign of chimerism.

This case, examined and published by the same research team involved in its investigation, offers one of the clearest and most direct illustrations available of a theme running throughout this book: chimerism, in a meaningful number of cases, produces detectable physical signs, sometimes quite visible ones, that go unrecognized for years or decades simply because chimerism sits so far outside the standard diagnostic vocabulary that most physicians, and most people generally, carry with them through ordinary life. It also offers a striking, specific, and scientifically grounded estimate regarding just how common the underlying biological starting point for chimerism, a pregnancy that begins with more than one embryo, may actually be. Researchers examining this case cited existing research suggesting that roughly one in eight single childbirths may have originally begun as a multiple pregnancy, a figure broadly consistent with, though somewhat higher than, the fifteen to thirty six percent range for vanishing twin syndrome specifically discussed in an earlier chapter, and reflecting the reality that different research methodologies, examining different populations and different stages of pregnancy, have produced a range of overlapping but not perfectly identical estimates for how frequently human pregnancies begin as multiples and resolve into singleton births.

If a meaningful fraction of these vanished twin pregnancies involve some degree of cell fusion or persistent genetic material exchange, rather than a clean and total loss with no lasting trace, then the pool of individuals walking around with some undiagnosed degree of chimerism could be substantially larger than the small number of confirmed, published cases would suggest on its own. This is not a fringe or speculative claim invented for the purposes of this book. It reflects the considered, published view of the actual research scientists who have investigated specific confirmed chimerism cases and who have direct professional experience with how difficult and how dependent on accident or unusual circumstance any individual chimerism diagnosis tends to be.

Why does chimerism remain so persistently underdiagnosed, given that the underlying biological starting point, twin pregnancies that resolve to a single birth, appears to be genuinely common rather than rare? Several distinct factors combine to produce this diagnostic gap, and understanding each of them clarifies why chimerism functions, in a real and specific sense, as a hidden phenomenon rather than merely a rare one.

The first and most fundamental factor is that there is, under ordinary circumstances, no medical reason to test for chimerism at all. Routine physical examinations, standard blood work, and general medical care of essentially every kind do not involve comparing genetic material sampled from multiple different tissues within the same patient's body, which is the only reliable method for detecting most forms of chimerism. A physician examining an ordinary patient has no clinical reason to suspect that the patient's blood might carry a different genetic profile than the patient's skin or reproductive tissue, and no standard diagnostic protocol exists that would prompt such a comparison in the absence of some unusual triggering circumstance, such as a failed organ transplant compatibility test, a legal paternity or maternity dispute, or a visibly striking physical anomaly like Taylor Muhl's divided skin pigmentation or the Washington father's two toned patches.

The second factor involves the specific and genuinely unpredictable way chimerism distributes itself throughout an affected individual's body. Because the original embryonic fusion occurs so early in development, before differentiated tissue layers have formed, there is no consistent, predictable pattern governing which of an individual's tissues will end up derived predominantly from one original embryo's cell line versus the other. In some documented cases, this distribution produces visible external signs. In many others, based on the working assumptions of researchers who study the field, it very likely does not, meaning the chimeric individual's skin, hair, and outward appearance offer no clue whatsoever to what a genetic comparison of different internal tissues might reveal.

The third factor is a genuine and reasonable diagnostic caution on the part of practicing physicians. When a genetic test produces a confusing or seemingly impossible result, the overwhelmingly most likely explanation, in the vast majority of real world cases, is simple laboratory error: sample mislabeling, cross contamination, or a technical processing mistake, rather than an exceedingly rare biological condition that most physicians will never encounter in their careers. This is, in fact, the medically appropriate first assumption in most circumstances, since laboratory errors, while themselves relatively uncommon in accredited testing facilities, remain considerably more common than confirmed tetragametic chimerism. The result is that a confusing genetic test finding is far more likely to be resolved through a request for retesting, which, if it also comes back with the same confusing result, may then prompt further investigation, than through an immediate leap to considering chimerism, unless something about the specific circumstances, as happened for both Lydia Fairchild's attorney and the investigators in the Washington paternity case, prompts a more thorough and sustained search for an explanation.

Given these combined obstacles to diagnosis, it is worth asking directly what, if anything, might change going forward, and here the growing popularity and increasing genetic sophistication of consumer ancestry and genealogy testing services offers a genuinely interesting and only recently emerging answer. As the Washington paternity case demonstrates directly, consumer genetic testing services, designed originally for ancestry research and extended family discovery rather than clinical diagnosis, examine a vastly larger number of genetic markers across the genome than traditional paternity or maternity testing, giving them meaningfully greater statistical power to detect the kind of subtle genetic irregularities that indicate chimerism, even in cases where the chimerism might not be severe or extensive enough to produce a complete mismatch on a more limited, traditional test panel.

Geneticists who have studied and published on this specific case have suggested that paternity and maternity testing protocols more broadly might benefit from incorporating additional genetic markers, similar in scope to those used by consumer ancestry services, specifically to improve detection of chimerism in future cases before they escalate into the kind of legal or personal crisis experienced by Lydia Fairchild and by the Washington father discussed in this chapter. Some researchers have gone further, proposing that laboratories accredited to perform legal paternity and maternity testing adopt a standardized secondary review protocol, triggered automatically whenever an initial result falls outside expected statistical ranges in specific, well characterized ways, rather than relying entirely on individual case workers or attorneys to notice an anomaly and pursue further investigation on their own initiative, as happened only through unusual persistence in both Lydia Fairchild's case and the Washington paternity case examined in this chapter. At the same time, researchers directly involved in this area of study have offered an important caution against overinterpreting consumer ancestry results as a reliable, standalone diagnostic tool for chimerism specifically, noting that these tests were designed and validated for a genuinely different purpose, genealogical relationship mapping, and that a confirmed chimerism diagnosis still requires the kind of careful, multi tissue genetic analysis and clinical interpretation that only specialized genetics laboratories and trained genetic counselors are currently equipped to provide reliably.

This growing intersection between consumer genetic testing and incidental chimerism discovery represents a genuinely new development in the broader history of the condition, one that did not exist in anything like its current form even in the years immediately following Lydia Fairchild's case. Millions of people worldwide have now used consumer ancestry testing services for reasons having nothing whatsoever to do with suspected chimerism, simple curiosity about ethnic heritage, interest in connecting with unknown extended family members, or participation in broader genealogical research projects. As the genetic sensitivity and sophistication of these consumer services continues to improve, and as the pool of people who have used them continues to grow into the tens of millions, it seems reasonably likely that additional incidental chimerism discoveries, following broadly the same pattern as the Washington paternity case, will continue to surface with somewhat greater frequency than the historical rate of documented cases might suggest, simply because more people than ever before are now undergoing the kind of detailed genetic testing capable of revealing the condition, even when chimerism detection was never the intended purpose of that testing in the first place.

The commercial genetic testing industry itself has, somewhat unexpectedly, become an informal but genuinely significant contributor to chimerism research more broadly, a role its founders and early customers could scarcely have anticipated when these companies were first established primarily to satisfy consumer curiosity about ancestry and distant family connections. Several of the larger consumer genetic testing companies maintain research divisions that, with appropriate customer consent protocols in place, analyze aggregated and anonymized data drawn from their enormous customer databases to identify broader genetic patterns and rare conditions, including instances of chimerism that surface incidentally within this aggregated data. This represents a genuinely novel research pathway that did not exist in any comparable form before the emergence of large scale consumer genetic testing, offering researchers access to a vastly larger pool of genetic data than any single academic research study, however well funded, could realistically hope to assemble on its own, and correspondingly offering a meaningfully improved, if still incomplete, opportunity to identify chimerism cases that would otherwise have gone entirely undetected throughout the affected individual's life.

This slow, incidental expansion of chimerism awareness, driven less by any coordinated medical research initiative and more by the unrelated commercial growth of consumer genetic testing, offers a genuinely hopeful, if unplanned, pathway toward finally closing at least part of the persistent gap between the likely true prevalence of human chimerism and the still remarkably small number of cases formally documented in the medical literature. It will not, on its own, identify every undiagnosed chimeric individual living today, the overwhelming majority of whom will likely go through their entire lives with no reason or occasion to undergo genetic testing detailed enough to reveal their condition. But it represents genuine, measurable progress in a field that, for most of its documented history, has depended almost entirely on individual, isolated instances of medical or legal crisis to bring any given case to light at all.

The Washington paternity case also raises a genuinely important question that researchers and genetic counselors specializing in chimerism have only recently begun to address in a systematic way: what should happen, from a clinical and counseling perspective, once an individual receives a confirmed chimerism diagnosis, particularly one arising incidentally through consumer testing rather than through a dedicated clinical genetics workup. Traditional genetic counseling protocols, developed primarily around far more common conditions such as inherited disease risk or carrier status for specific genetic disorders, were not originally designed with chimerism specifically in mind, given how rarely the condition has historically been diagnosed. Genetic counselors who have worked directly with confirmed chimeric patients have generally adapted existing counseling frameworks to address the specific concerns chimerism tends to raise, drawing heavily on the personal testimony of previously diagnosed individuals like Taylor Muhl to help reassure newly diagnosed patients that chimerism, however strange and rare its explanation, does not typically require ongoing medical management or intervention beyond addressing whatever specific triggering health concern originally prompted the investigation.

This counseling gap is likely to narrow over time as chimerism diagnoses, while remaining rare in absolute terms, continue to become somewhat more frequent through the combination of factors discussed throughout this chapter and the chapters preceding it: growing consumer genetic testing adoption, improving detection sensitivity, and a slowly expanding base of published case literature that future physicians and genetic counselors will have available to draw upon, in exactly the way Karen Keegan's published case became available to Lydia Fairchild's attorney at precisely the moment he needed it most.

The next chapter looks more directly forward, examining how the broader landscape of genetic testing, forensic science, and reproductive technology is likely to continue evolving in ways that will shape how chimerism, and the deeper questions about genetic identity it raises, are understood and encountered in the years ahead.

CHAPTER 13

Chimerism and the Future of Genetic Testing

Genetic testing has become, over the span of little more than a single generation, dramatically cheaper, dramatically faster, and dramatically more embedded in ordinary life than almost anyone working in the field could have confidently predicted when the first draft of the complete human genome was published in the early years of this century. Sequencing an entire human genome, a task that once required years of coordinated international effort and billions of dollars in combined funding, can now be accomplished for a small fraction of that cost in a matter of days. Consumer genetic testing kits, ordered online and processed from a simple saliva sample mailed to a laboratory, have become commonplace enough that tens of millions of people worldwide have now used them for reasons ranging from serious health screening to casual curiosity about ancestry. Prenatal genetic testing, once limited to invasive procedures carrying real risk to the pregnancy, can now screen for a range of chromosomal conditions using nothing more than a blood draw from the pregnant mother, relying directly on the small fragments of fetal DNA that cross the placental barrier during pregnancy, the same biological process responsible for the microchimerism examined in an earlier chapter.

This rapid decline in sequencing cost deserves a brief moment of concrete illustration, because the scale of the change is easy to state abstractly and easy to underestimate without a specific point of comparison. Early efforts to sequence a single complete human genome, undertaken as part of the international collaborative project that produced the first full human genome sequence, required roughly a decade of sustained international research effort and a combined investment reaching into the billions of dollars. Sequencing technology has since improved so dramatically that a genome of comparable completeness can now be sequenced for a cost representing a tiny fraction of that original figure, within a timeframe measured in days rather than years, a rate of technological improvement that has, by some comparisons drawn by researchers who study the pace of change in scientific instrumentation, outpaced even the well known rapid historical improvement curve associated with computer processing power.

This rapid and continuing expansion of genetic testing technology carries direct and significant implications for how chimerism specifically is likely to be understood, detected, and managed in the years ahead, extending well beyond the incidental consumer ancestry testing discoveries discussed in the previous chapter.

Forensic science, the field most directly and repeatedly confronted with the practical consequences of chimerism throughout the cases examined earlier in this book, is actively evolving its own standard practices in response to accumulating awareness of the phenomenon. Forensic genetics researchers who have published specifically on chimerism's implications for criminal investigation and human identification have increasingly recommended that forensic laboratories develop and adopt standardized protocols for handling cases involving known or suspected chimerism, whether naturally occurring or resulting from a documented bone marrow transplant. These recommendations include maintaining careful medical history intake procedures capable of flagging prior transplant recipients before genetic samples are processed, developing validated methods for analyzing multiple tissue types from a single individual when initial results appear anomalous or contradictory, and building broader professional awareness among practicing forensic scientists regarding the range of documented chimerism related complications, so that a confusing genetic profile prompts appropriate further investigation rather than being dismissed as simple laboratory contamination or, worse, being accepted uncritically as reliable evidence in a criminal or civil legal proceeding.

Genetic genealogy, the increasingly influential investigative technique that combines DNA analysis with traditional genealogical research to identify unknown suspects or unidentified human remains, represents a particularly active area of ongoing research specifically regarding chimerism's potential to complicate results. This technique has grown dramatically in both capability and public prominence in recent years, credited with helping resolve a number of previously unsolvable cold cases by identifying suspects through distant genetic relatives found in consumer genealogy databases rather than through a direct genetic match to the suspect themselves. Because this technique relies on detecting genetic relationships across an extended family tree rather than simply confirming or excluding a single direct match, researchers examining how chimerism might specifically affect genetic genealogy work have noted that a chimeric individual's genetic profile could, in principle, generate confusing or even actively misleading family tree relationships, an area of forensic science methodology that remains under active study as the broader field continues to mature and formalize its investigative standards.

Reproductive medicine and assisted reproductive technology represent another domain in which chimerism related research is likely to continue expanding, driven partly by the genuinely documented cases of monochorionic dizygotic twin pregnancies and their associated blood chimerism risk discussed in an earlier chapter, several of which specifically involved pregnancies conceived through intracytoplasmic sperm injection and other assisted reproductive procedures. As assisted reproductive technology continues to become more widely used worldwide, and as fertility clinics continue refining their embryo transfer and implantation procedures, researchers in this field have suggested that better understanding of the specific circumstances under which dizygotic twin embryos might develop an unexpected shared placental connection, producing blood chimerism risk that would not arise in an ordinary, separately implanted dizygotic twin pregnancy, could help inform future clinical protocols aimed at minimizing complications like twin to twin transfusion syndrome in these less common but clinically significant pregnancy scenarios.

The organ transplantation field, and the closely related emerging field of laboratory grown interspecies chimeric organs examined in an earlier chapter, continues to advance along a trajectory that will likely make deliberately engineered chimerism an increasingly familiar and increasingly clinically significant medical reality over the coming years, independent of the naturally occurring forms of chimerism this book has primarily focused on. As gene editing technology continues to improve in both precision and affordability, and as researchers continue refining the blastocyst complementation techniques capable of growing human relevant organs inside animal hosts, the medical and ethical questions surrounding deliberately created human animal chimeras seem likely to move from a relatively specialized area of biomedical research discussion toward a more prominent position in broader public conversation, particularly if and when this research approach progresses from its current experimental stage toward genuine clinical application capable of meaningfully addressing the persistent shortage of transplantable human organs.

Beyond these specific applied fields, the continuing decline in the cost of comprehensive genetic sequencing carries a more diffuse but potentially quite significant implication for chimerism research generally: the increasing feasibility of large scale population studies specifically designed to detect chimerism at a level of rigor and scale that has simply never been logistically or financially achievable before. Nearly every documented case of human chimerism examined throughout this book was identified incidentally, discovered only because some unrelated circumstance, a legal dispute, a failed transplant compatibility test, a visible physical anomaly, or a consumer ancestry test, happened to trigger genetic testing sensitive enough to reveal the underlying condition. No large scale, systematic research study has yet been conducted specifically designed to screen a substantial general population sample for chimerism using consistent methodology across every participant, in significant part because the cost and logistical complexity of comparing genetic material across multiple tissue types for a genuinely large sample of research subjects has, until relatively recently, remained prohibitively expensive and impractical to justify for a condition assumed, based on the small number of documented cases, to be exceedingly rare.

International collaboration among genetics researchers studying rare conditions has also expanded considerably in recent years, offering a further pathway through which chimerism research specifically stands to benefit going forward. Rare disease research more broadly has historically suffered from a basic statistical problem: any single research institution, however well resourced, will typically encounter too few confirmed cases of an exceedingly rare condition to conduct statistically meaningful research on its own. International rare disease registries and collaborative research networks, increasingly common across many areas of medical genetics, allow researchers studying conditions like chimerism to pool case data and biological samples across multiple countries and institutions, substantially increasing the total number of confirmed cases available for combined study compared to what any individual research group could hope to assemble independently. Several such collaborative registries already exist for a range of rare genetic conditions, and researchers specializing in chimerism have begun advocating for a dedicated, formally organized international chimerism registry specifically designed to aggregate case data, biological samples, and long term outcome information from confirmed chimeric individuals worldwide who consent to participate, a resource that, if successfully established and maintained over time, could meaningfully accelerate the field's currently slow, case by case pace of discovery.

As sequencing costs continue their sustained multi decade decline, and as automated laboratory techniques for comparing genetic material across multiple tissue samples become increasingly efficient and increasingly affordable, the prospect of a genuinely comprehensive, systematically designed population level study of chimerism prevalence becomes correspondingly more realistic. Such a study, were it conducted with appropriate scientific rigor across a sufficiently large and representative population sample, could finally provide researchers with something the field has lacked throughout its entire documented history: a genuinely reliable, empirically grounded estimate of how common chimerism actually is in the general population, rather than the current situation, in which researchers must rely on educated extrapolation from indirect evidence, such as vanishing twin syndrome prevalence rates and the handful of incidentally discovered confirmed cases, to construct necessarily uncertain estimates of the true scope of the phenomenon.

Artificial intelligence and machine learning tools, increasingly deployed across many areas of genetic research and clinical diagnostics, may also play a growing role in chimerism detection specifically, particularly in helping flag genetic testing results that show subtle statistical irregularities consistent with chimerism but not severe enough to produce an obvious, unmistakable mismatch of the kind that eventually prompted investigation in cases like Lydia Fairchild's or the Washington paternity case. Pattern recognition tools capable of automatically screening large volumes of genetic testing data for these subtler statistical signatures, work that would be prohibitively time consuming and expensive for human geneticists to conduct manually across every test processed by a busy commercial or clinical laboratory, could in principle help identify a meaningfully larger number of chimerism cases going forward than the current system, which depends almost entirely on either an obvious and dramatic mismatch or a human geneticist's individual curiosity and persistence, has managed to identify throughout its documented history to date.

None of these emerging developments, taken individually or collectively, are likely to transform chimerism from a rare and specialized area of genetic research into a routinely diagnosed medical condition encountered regularly in ordinary clinical practice. The fundamental biological rarity of the more dramatic tetragametic form of chimerism examined throughout much of this book, along with the practical and financial barriers that continue to limit how frequently and how thoroughly any given individual's genetic material actually gets tested across multiple distinct tissue types, will likely continue to keep chimerism detection heavily dependent on some form of triggering circumstance for the foreseeable future, rather than becoming a matter of routine screening built into standard medical care. But the cumulative direction of change across forensic science, reproductive medicine, organ transplantation research, population genetics, and diagnostic technology all point toward a future in which chimerism, while remaining genuinely rare in its most dramatic documented forms, becomes an increasingly well understood, increasingly well anticipated, and increasingly less mysterious feature of the broader genetic landscape that medicine, law, and science must navigate.

This matters beyond the relatively small number of individuals who will ever receive a confirmed chimerism diagnosis themselves, because the underlying lesson chimerism teaches, that DNA testing, for all its genuine scientific power and its thoroughly deserved reputation for reliability in the overwhelming majority of ordinary cases, is not infallible and cannot always be treated as a simple, context free, unambiguous fact, carries implications that extend into essentially every area of modern life now touched by genetic testing, from criminal justice to family law to reproductive medicine to basic scientific research. A society that better understands the genuine, if statistically rare, edge cases where genetic testing can mislead rather than clarify, is a society better equipped to use this genuinely powerful technology wisely, with appropriate confidence in its ordinary reliability balanced against appropriate humility regarding its occasional and well documented limitations.

The legal profession specifically stands to benefit from continuing education initiatives that incorporate chimerism related case law and forensic research into standard training for attorneys who regularly handle family law, immigration, and criminal defense matters involving genetic evidence. Alan Tindell's successful defense of Lydia Fairchild depended heavily on his own individual persistence and a fortunate research discovery made under significant time pressure, a pattern that, however inspiring as an individual story of advocacy, does not represent a reliable or scalable system for ensuring that future attorneys facing similarly confusing genetic evidence will know where to look or what questions to ask. Law schools and continuing legal education programs that address forensic genetics have increasingly begun incorporating chimerism as a specific topic within broader coursework on DNA evidence, reflecting a growing professional recognition that the kind of specialized knowledge Tindell had to assemble largely on his own, under genuine time pressure and with his client's custody of her children hanging in the balance, should ideally be available as part of standard professional training long before any individual attorney encounters a case requiring it.

The final chapter of this book steps back from these more technical and forward looking considerations to offer a closing reflection on what chimerism, taken as a whole, reveals about the nature of individual identity, biological inheritance, and the surprisingly permeable boundaries that separate one human being from another, boundaries this book has shown to be considerably less absolute, in a meaningful number of well documented cases, than most people ever have occasion to realize.

CHAPTER 14

What the Vanishing Twin Leaves Behind

Every person reading this book began, in the most literal biological sense, as a single cell. That cell divided, and divided again, and kept dividing, building itself into a brain capable of holding a book's worth of ideas, a heart that has beaten without pause since sometime before birth, hands capable of turning a page. Somewhere in that long, unbroken chain of cellular division, for the overwhelming majority of readers, nothing unusual happened at all. One zygote became one body, uniformly, completely, and the story of that body's genetic identity is as simple and as settled as such stories generally are.

For a small, largely unmeasured, and mostly undiscovered minority of people, something else happened instead. Two separate beginnings, two separate zygotes, each carrying its own complete genetic blueprint, met somewhere in the earliest days of embryonic development and did not stay separate. They became, together, one continuing story, one body, one life, built from what would otherwise have been two.

This book has moved through the many different shapes that story can take. It began with Lydia Fairchild, nearly losing her own children to a bureaucracy that trusted a single genetic test more than it trusted a mother's own unwavering certainty about who she was. It moved through Karen Keegan, whose thyroid tissue quietly held the genetic key to a mystery her blood alone could never have solved. It paused with Taylor Muhl, who spent decades believing an ordinary birthmark explained what turned out to be the visible signature of a twin she had carried inside her own skin since before she was born. It examined the intimate, universal chimerism of pregnancy itself, in which every mother and every child who has ever shared a womb carries forward, for years and often decades afterward, small living traces of each other, a form of chimerism so common that it touches nearly every family, and so quiet that almost none of those families ever think to notice it. It traced the accidents of shared placental blood supply that can leave twin siblings, otherwise entirely separate people, carrying a permanent trace of one another circulating in their blood. It followed chimerism into courtrooms and crime laboratories, where a single, unexamined assumption about the reliability of DNA evidence nearly cost an innocent mother her family and did, in at least one documented case, briefly implicate an innocent man serving time for a crime he did not commit. It ventured into laboratories where scientists now deliberately construct chimeras of their own, growing human tissue inside animal hosts in pursuit of solving one of modern medicine's most persistent and lethal shortages. It reached back nearly three thousand years to a monster stitched together from a lion, a goat, and a serpent, and found that the ancient imagination that named this monster had captured something about coexisting, complete identities that modern genetics would not confirm as literal biological fact until thousands of years later. And it sat, without pretending to resolve, in the genuinely difficult philosophical territory chimerism opens up regarding what actually makes any of us a single, continuous self.

What connects all of these threads, more than any specific shared mechanism or shared case, is a single underlying insight that chimerism, in each of its many forms, keeps returning to from a slightly different angle: the boundary between one person and another, which everyday experience treats as obvious, absolute, and essentially beyond question, is, at the cellular and genetic level, considerably less fixed than it appears. Not fixed in the dramatic, identity dissolving way that popular, sensationalized coverage of chimerism sometimes implies, breathlessly suggesting that chimeric individuals are somehow not fully themselves, or are secretly two people trapped in one body. The evidence and the testimony examined throughout this book point toward something quieter and, in its own way, more remarkable: people who are entirely, unambiguously, continuously themselves, experiencing the same unified sense of identity that every other person experiences, while also, at the level of their actual physical cells, carrying more genetic history and more genetic company than most of us ever imagine possible.

This is worth sitting with for a moment, because it reframes what might otherwise seem like a purely medical curiosity into something closer to a genuinely humbling reminder about the limits of our everyday intuitions regarding our own bodies. Most people go through their entire lives assuming, without ever having reason to question it, that their body is genetically uniform from scalp to fingertip, one continuous story with one continuous author. For the overwhelming majority of people, that assumption holds, and holds completely. But it holds as a statistical reality rather than a biological law, a very strong tendency rather than an absolute guarantee, and the exceptions, rare as they individually are, occur with sufficient collective frequency, once every form of chimerism examined in this book, from tetragametic fusion to blood chimerism to the near universal microchimerism of pregnancy, is properly accounted for, that they deserve to be understood as a genuine feature of human biology rather than dismissed as a mere medical footnote.

There is, woven through the specific stories this book has told, a recurring and genuinely moving pattern worth naming directly: chimerism, again and again, turns out to mean that the people we love, and the people who came before us, and the twins we may never have known we almost had, remain with us more literally, more physically, than ordinary intuition would ever suggest. A mother carries her child's cells, quietly, for decades after that child has grown and moved into an independent adult life of their own. A child carries their mother's cells in equal measure, a permanent biological inheritance passed down in a form far more direct than genes alone. A person born as one of what would have been fraternal twins may carry, distributed unevenly throughout organs they will never think to test, the living genetic legacy of a sibling who never took an independent breath, but who did not, in the fullest biological sense, simply disappear either.

There is a particular kind of humility this book hopes to have modeled throughout, one worth naming explicitly as this closing chapter draws toward its end. Every case examined across these pages, from Lydia Fairchild's custody battle to Karen Keegan's kidney crisis to Taylor Muhl's divided skin to the Washington father's unexpected paternity result, began with someone, a caseworker, a surgical team, an attorney, a genetic counselor, encountering a result that made no sense according to everything they thought they already understood, and choosing to keep looking rather than to force the confusing evidence into a familiar but ultimately incorrect explanation. That choice, to sit with genuine confusion long enough to find the actual answer rather than the convenient one, runs through this entire book as a kind of quiet, recurring heroism, exercised not by any single dramatic actor but by an accumulated chain of careful, curious people across medicine, law, and science, each contributing one piece of a puzzle that no single person could have solved alone.

None of this should be read as offering false comfort regarding the genuine grief associated with pregnancy loss, vanishing twin syndrome, or miscarriage, subjects this book has treated with the seriousness they deserve rather than as occasions for uplifting reframing. The loss involved in these experiences is real, and the biological fact that some trace of an unborn twin or an early pregnancy may persist within a parent's or a sibling's body afterward does not undo or diminish that loss in any way that should be pressed upon someone grieving it. But for those who do find some measure of meaning or comfort in learning of these biological realities, whether a parent who has experienced pregnancy loss, an adult who discovers in later life that they began as one of a twin pregnancy, or simply a curious reader encountering this science for the first time, that meaning belongs to them to find in their own way, and the science, at minimum, does not foreclose it.

Chimerism also offers, in its own quiet way, a genuinely useful corrective to a habit of thought that extends well beyond genetics specifically: the tendency to treat categories that feel intuitively obvious and self evident, one person, one body, one continuous genetic identity, as though they were logically necessary rather than simply, overwhelmingly, the common case. Biology is full of categories that hold true with remarkable consistency and yet admit genuine, well documented exceptions once examined closely enough and broadly enough, and the appropriate response to discovering such an exception is neither to panic that the entire category has collapsed, nor to dismiss the exception as insignificant simply because it is rare. The appropriate response, modeled throughout the careful, methodical work of every researcher and physician whose investigations this book has drawn upon, from the team that identified Karen Keegan's thyroid tissue mismatch to the geneticists who traced a Washington man's confusing paternity result back to an unborn twin, is patient, rigorous curiosity: a willingness to keep testing, keep questioning, and keep looking for the actual underlying explanation, even when, and perhaps especially when, the answer turns out to be something as genuinely strange as discovering that a body assumed to be one person is, biologically speaking, more accurately understood as two.

Lydia Fairchild kept her children because a lawyer refused to accept an official, seemingly authoritative answer that did not match the reality he could see standing in front of him, and because a small community of researchers and physicians had done the careful, patient work years earlier, with an entirely different patient and an entirely different medical crisis, to document and publish a phenomenon rare enough that most of their colleagues would never personally encounter it. That published research, sitting quietly in a medical journal until the moment someone needed it most urgently, saved a family that a state bureaucracy had been prepared to tear apart on the strength of a single blood sample.

It is worth remembering, in closing, that every individual profiled across this book eventually returned to an ordinary life once their particular chimerism related crisis or discovery had passed. Lydia Fairchild raised her children. Karen Keegan recovered from her transplant. Taylor Muhl continues her career as a performer and model while advocating publicly for greater chimerism awareness. The Washington father profiled in a later chapter continues raising his sons, both of them, regardless of which genetic lineage happened to produce each child's particular set of chromosomes. None of these individuals were transformed into someone fundamentally different by their diagnosis. Each simply gained a piece of accurate information about their own biology that had, until the moment of discovery, remained hidden, and each incorporated that information into a life that had already been, and remained afterward, entirely and recognizably their own.

This is, in the end, perhaps the most important practical takeaway this book can offer any reader who finishes it wondering what to make of everything chimerism represents. Rare biological realities, however strange or however dramatic the language used to describe them in a headline or a book title, do not usually announce themselves as ruptures in the ordinary business of living. They tend, instead, to arrive quietly, discovered incidentally, absorbed gradually, and folded eventually into a life that continues largely as it always has, simply carrying one additional, previously unknown fact about the particular, unrepeatable biological history that produced it. The twin who wasn't, wherever and however she persists across the many forms this book has traced, does not need to be feared, mourned indefinitely, or treated as evidence of some deeper fracture in what it means to be a single person. She is, more often than not, simply part of the story, carried quietly and without incident by someone who was always, from their own first person perspective, exactly who they always knew themselves to be.

That is, in the end, as good a case as any for why the strange, rare, quietly humbling biology examined throughout this book is worth understanding, worth documenting, and worth continuing to research, even for a phenomenon that will likely never touch the daily life of the overwhelming majority of people who read about it. Somewhere, right now, another Lydia Fairchild may be sitting across from another caseworker, receiving another impossible seeming result from another routine test, wondering how a piece of scientific paper could presume to contradict everything she has always known to be true about herself and about the family she built. The answer, when it finally comes, if it comes at all, will very likely trace back to exactly this kind of patient, careful, previously published science, sitting ready and waiting in exactly the way Karen Keegan's case was ready and waiting for Lydia's attorney, for the moment someone needs it enough to go looking.

Readers who have followed this book from its opening chapter to this closing one have, in effect, been given a small piece of exactly that kind of readiness. Should a friend, a family member, or a reader's own future genetic test ever produce a result that seems to defy explanation, a mismatch too strange and too confident to make ordinary sense, this book's central lesson is worth carrying forward: the strangeness may not be an error at all. It may be biology working exactly as it should, along a rarer and more surprising path than most people ever have occasion to learn about, one that ancient storytellers reached for with a lion, a goat, and a serpent, and that modern genetics has finally, carefully, and only very partially, learned to explain in full.

The twin who wasn't, in each of the many forms examined across this book, very often turns out not to have entirely vanished after all. She persists, quietly, in blood and thyroid tissue and reproductive cells and skin pigmentation and, more universally and more gently than any of the dramatic individual cases this book has examined, in the cells every mother and every child exchange and carry forward, largely unnoticed, for the whole of their lives together. The boundary between one person and another, it turns out, is not quite as solid as it looks. And knowing that, carefully and accurately rather than sensationally, makes it a little easier to appreciate just how much more connected, and how much more biologically generous toward one another, our bodies already are than most of us ever have reason to realize.

www.ingramcontent.com/pod-product-compliance
Lightning Source LLC
La Vergne TN
LVHW100143061126
851857LV00010B/216
9798235626997